Lecture Notes in Computer Sc

Commenced Publication in 1973
Founding and Former Series Editors:
Gerhard Goos, Juris Hartmanis, and Jan van Leeuwen

T0238218

Editorial Board

Babak Falsafi   T.N. Vijaykumar (Eds.)

Power-Aware Computer Systems

4th International Workshop, PACS 2004
Portland, OR, USA, December 5, 2004
Revised Selected Papers

 Springer

Volume Editors

Babak Falsafi
Carnegie Mellon University
Electrical and Computer Engineering Dept.
A305 Hamerschlag Hall, 5000 Forbes Avenue, Pittsburgh, PA 15213, USA
E-mail: babak@cmu.edu

T.N. Vijaykumar
Purdue University
School of Electrical and Computer Engineering
Department of Computer Science
ECE/EE 465 Northwestern Avenue, West Lafayette, Indiana 47907-1285, USA
E-mail: vijay@ecn.purdue.edu

Library of Congress Control Number: 2005936777

CR Subject Classification (1998): B.7, B.8, C.1, C.2, C.3, C.4, D.4

ISSN   0302-9743
ISBN-10   3-540-29790-1 Springer Berlin Heidelberg New York
ISBN-13   978-3-540-29790-1 Springer Berlin Heidelberg New York

Springer is a part of Springer Science+Business Media

springeronline.com

© Springer-Verlag Berlin Heidelberg 2005

Typesetting: Camera-ready by author, data conversion by Scientific Publishing Services, Chennai, India
Printed on acid-free paper   SPIN: 11574859   06/3142   5 4 3 2 1 0

Preface

Welcome to the proceedings of the Power-Aware Computer Systems (PACS 2004) workshop held in conjunction with the 37th Annual International Symposium on Microarchitecture (MICRO-37). The continued increase of power and energy dissipation in computer systems has resulted in higher cost, lower reliability, and reduced battery life in portable systems. Consequently, power and energy have become first-class constraints at all layers of modern computer systems. PACS 2004 is the fourth workshop in its series to explore techniques to reduce power and energy at all levels of computer systems and brings together academic and industry researchers.

The papers in these proceedings span a wide spectrum of areas in power-aware systems. We have grouped the papers into the following categories: (1) microarchitecture- and circuit-level techniques, (2) power-aware memory and interconnect systems, and (3) frequency- and voltage-scaling techniques.

The first paper in the microarchitecture group proposes banking and write-back filtering to reduce register file power. The second paper in this group optimizes both delay and power of the issue queue by packing two instructions in each issue queue entry and by memorizing upper-order bits of the wake-up tag. The third paper proposes bit slicing the datapath to exploit narrow width operations, and the last paper proposes to migrate application threads from one core to another in a multi-core chip to address thermal problems.

The second group of papers on power-aware memory and interconnects starts with a contribution which proposes hardware–software co-operation to reduce main memory power dissipation. The paper suggests combining process-level information from the software and DRAM-bank-level information from the hardware for significant power reduction. The second paper in this group uses compiler-assist to make hardware prefetching more energy efficient by filtering out unnecessary and ineffective prefetching. The third paper explores modeling of external bus power dissipation and evaluation of coding techniques for bus power reduction. The last paper proposes context-independent coding for reducing power in off-chip interconnects to avoid the disadvantage of context-dependent coding not being applicable to commodity memories because of requiring collaboration between the memory controller and SDRAM.

The last group proposes frequency- and voltage-scaling techniques. The first paper in this group recommends throttling processor clock speed during low-utilization phases. The second paper scales the processor voltage according to the CPU-boundedness of the application. The third paper investigates the potential of hardware overprovisioning to increase throughput in data centers while remaining within a power budget. The last paper shows a detailed breakdown of power consumption in the various components of a modern laptop.

The success of PACS 2004 has been due to the high quality of the submissions, the efforts of the Program Committee, and the attendees. We would like to thank Vivek De for his informative keynote address, which described design challenges and opportunities for power-limited microprocessors. We would also like to thank Jose Gonzalez, Glen Reinman, Srikanth Srinivasan and other members of the MICRO-37 Organizing Committee who helped arrange the local accommodation and publicize the workshop.

December 2004   Babak Falsafi
                                                    T. N. Vijaykumar

PACS 2004 Program Committee

Babak Falsafi, *Carnegie Mellon University (Co-chair)*
T. N. Vijaykumar, *Purdue University (Co-chair)*

David Albonesi, *Cornell University*
Csaba Andras Mortiz, *University of Massuchusetts*
Krste Asanovic, *Massachusetts Institute of Technology*
Luca Benini, *Universitá di Bologna*
Frederic Chong, *University of California, Davis*
Kanad Ghose, *State University of New York, Binghamton*
Christoforos Kozyrakis, *Stanford University*
Uli Kremer, *Rutgers University*
Charles Lefurgy, *IBM, Austin Research Lab*
Yung-Hsiang Lu, *Purdue University*
Avi Mendelson, *Intel, Israel Microarchitecture Lab*
Andreas Moshovos, *University of Toronto*
Daniel Mosse, *University of Pittsburgh*
Vijaykrishnan Narayanan, *Pennsylvania State University*
Li-Shiuan Peh, *Princeton University*
Parthasarathy Ranganathan, *HP Labs*
Eric Rotenberg, *North Carolina State University*
Mircea Stan, *University of Virginia*
Se-Hyun Yang, *Samsung*

Table of Contents

An Optimized Front-End Physical Register File with Banking and Writeback Filtering

Miquel Pericàs[1,3], Ruben Gonzalez[1], Adrian Cristal[1],
Alex Veidenbaum[2], and Mateo Valero[1,3]

[1] Computer Architecture Department, Technical University of Catalonia (UPC)
[2] Information and Computer Science, University of California at Irvine (UCI)
[3] Barcelona Supercomputing Center (BSC)
{mpericas, gonzalez, adrian, mateo}@ac.upc.edu,
alexv@matrix.ics.uci.edu

Abstract. Register file design is one of the critical issues facing designers of out–of–order processors. Scaling up its size and number of ports with issue width and instruction window size is difficult in terms of both performance and power consumption. Two types of register file architectures have been proposed in the past: a future logical file and a centralized physical file.

The centralized register file does not scale well but allows fast branch mis–prediction recovery. The Future File scales well, but requires reservation stations and has slow mis–prediction recovery. This paper proposes a register file architecture that combines the best features of both approaches. The new register file has the large size of the centralized file and its ability to quickly recover from branch misprediction. It has the advantage of the future file in that it is accessed in the "front end" allowing about 1/3rd of the source operands that are ready when an instruction enters the window to be read immediately. The remaining operands come from bypass logic / instruction queues and do not require register file access. The new architecture does require reservation stations for operand storage and it investigates two approaches in terms of power–efficiency.

Another advantage of the new architecture is that banking is much easier to use in this case as compared to the centralized register file. Banking further improves the scalability of the new architecture. A technique for early release of short–lived registers called *writeback filtering* is used in combination with banking to further improve the new architecture. The use of a large front–end register file results in significant power savings and a slight IPC degradation (less than 1%). Overall, the resulting energy–delay product is lower than in previous proposals.

1 Introduction

Memory-based structures in the core of modern microprocessors have increasing energy requirements as frequencies grow. One such structure is the register file. Its size and the number of read/write ports required increases with issue width making it difficult to implement at high clock frequencies.

Two main approaches to register file design were used in the past, neither of which solved the above-mentioned problems. One approach was an architecture based on the Future file, which has a logical register file updated in commit and the future register

B. Falsafi and T.N. Vijaykumar (Eds.): PACS 2004, LNCS 3471, pp. 1–14, 2005.
© Springer-Verlag Berlin Heidelberg 2005

file in the "front–end" holding the most recent, uncommitted value for each logical register. The advantages of the future file are that it is not very large, has no renaming, can be read in the front–end and is not written if a more recent instruction assigning it is in the window. The disadvantages are that on branch mis–prediction, intermediate register values need to be recovered (typically after the mis–predicted branch commits), it needs reservation stations in the back-end, and its size cannot be increased. The mis–prediction recovery can lead to a significant IPC loss, especially given increasing memory latencies.

An alternative approach is a single, large physical register file, without a separate architectural register file. It is typically accessed after an instruction is scheduled to execute, even if source operand values were available when the instruction entered the window. This is the approach in the MIPS R10000 [1] and many later processors. Its advantages are increased size and fast mis–prediction recovery. Disadvantages are more complex renaming and longer value lifetime in the file due to lack of logical register file. Overall, it needs to be both large and heavily multi–ported, making it difficult to implement and increases its energy consumption significantly.

The new architecture proposed in this paper combines the best features of the two above–mentioned approaches: arbitrary size and fast mis–prediction recovery of the physical register file; and placement in the front–end, early operand read, and potential lack of write–back of the future file. It can be thought of as a physical register file moved to the front end and accessed after renaming. This allows a large fraction of operands to be accessed as an instruction enters the window, which is now the only read access to the register file. These values are stored in "reservation stations" integrated into the instruction queue, which can also be thought of as a replicated portion of the register file. A value coming from writeback may be written to this file if there are instructions waiting for it. Finally, many registers hold values for mis–prediction recovery, some of which can be released if they cannot affect recovery.

The approach proposed here uses a single register file containing all physical registers, the Front-end Physical Register File (FPRF). Thus restarting execution after a mispredicted branch can be done using a rename map recovery from check-points made on conditional branches.

As source operand registers are renamed, it can be determined if a register value has already been computed. The FPRF is read only in this case, significantly reducing its access frequency. Combined with the higher IPC due to faster branch recovery, it has a better energy-delay product compared to the two traditional approaches.

A new structure to hold such "early read" values is created in the instruction queue payload RAM. Its function is similar to that of reservation stations. It is smaller than the physical register file and thus consumes less energy. It is written into by completing instructions, if the produced value is a source operand of a waiting instruction.

This paper also investigate the use of banking in the FPRF architecture. Due to lower access frequency of the FPRF this is much easier to do than in a standard centralized physical register file Finally, writeback filtering, a technique to eliminate unnecessary writebacks into the register file is investigated and shown to be quite effective.

2   Related Work

The body of related work on register file energy optimization is large. Many recent papers have proposed mechanisms to reduce the number of the ports by means of modifying the register file architecture, such as [4] [5] [6] [7]. A reduced number of ports may be more efficient both in terms of energy and access time, which can improve performance.

A different approach is to reorganize the registers into several files, concentrating most activity on small files with low power consumption. [8] is an example of this approach based on the isolation of narrow operands. Hierarchical register files, such as those presented in [9] [10] [11] and clustering techniques such as [12] [5] are another example of this technique, which effectively trades size, speed and power consumption.

Another research direction has focused on changing the register allocation algorithm to reduce the register requirements of the architecture. *Early Release* frees registers before the commit stage of the next instruction that writes to the same logical registers [13] [14] [15]. *Virtual registers* [16] try to delay the allocation of the physical register until the writeback stage of the instruction. Another approach to reduce registers is to exploit repeated values in the registers [17] [18].

Our approach is somewhat based on the *Future File* organization which was proposed in 1984 [19]. In the original proposal, operands are provided to instructions via a logical register file in the front-end which received the name of *Future File*. The main difference with our architecture is that we are basing our design around the concept of physical registers to identify the state of the processor. Thus, while a Future File architecture can only recover from a mispredicted branch by draining the ROB, our proposed architecture can recover directly from the physical registers. Future File architectures are still being used in the form of the AMD K7 and K8 microarchitectures [20].

The future file can be extended with rename buffers to provide access to the full processor state at once. This has been implemented in the PPC620 [3] and POWER3 processors. However, these two processors still require the architectural state to be copied from the rename buffers during retirement. Having an architectural register file in the front-end shortens the pipeline one stage (access can be performed in parallel to rename stage) but increases the number of on-chip register transfers.

Research by Tseng et al. on banked register files [21] proposed an efficient implementation of banking for the register file of a MIPS R10000-like architecture. In the following sections is will be compared to the architecture proposed in this paper.

Finally, the *Writeback Filtering* technique, based on the release of short-lived values, is described in [22] and, in the context of VLIW architectures, in [23]. However, as will be shown here, the specific architectecture presented here allows to support *Writeback Filtering* in innovative ways.

3   Front-End Physical Register File

This section describes the *Front-End Physical Register File Architecture* (FPRF). The FPRF pipeline provides instructions with their operands as soon as the operands are available. Further, it implements a central physical register file in the front-end that

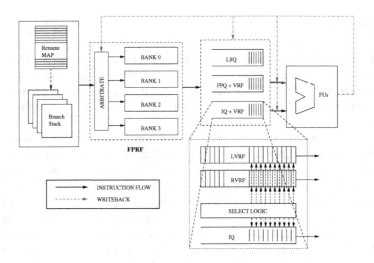

Fig. 1. The Front-End Physical Register File Architecture

allows for fast recovery with little complexity. It also allows to apply banking with high efficiency.

The FPRF Architecture, like a Future File, reads available registers in the front end. However, in this approach the registers are accessed via a mapping into a centralized register file that contains all registers. This has two implications:

1. Access to computed values in the front-end needs to be delayed until the rename stage has completed.
2. The number of registers in the front-end, being equal to the total number of registers, is much larger than it is in a Future File Architecture.

Figure 1 shows the FPRF architecture. Instructions, after going through the decode stage, enter the rename stage where source and destination registers are mapped to physical registers. Using this information an instruction may access the FPRF, a two stage process consisting of arbitration and data access. After available values have been given to the instruction, it is inserted into the corresponding instruction queue.

The back-end pipeline works as follows: When a functional units generates a result, the register tag is sent to all instructions in the queue. If there is an instruction waiting for it, the value is written into the corresponding entry of the Value Register File (VRF), which is part of the payload RAM of the instruction queue. The VRF is driven by the wakeup logic signals and can be implemented as a register file that does not require a decode stage. The value also gets written into the FPRF, as indicated by its physical register designator. There is also a possibility that a value is bypassed to a dependent instruction.

3.1   FPRF Pipeline

The pipeline of the FPRF Microarchitecture is shown in Fig. 2. It adds one stage to a commonly used 8-stage pipeline consisting of: fetch, decode, rename, queue, issue,

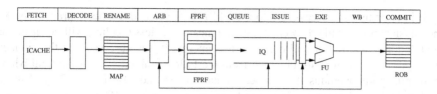

Fig. 2. The Pipeline of the FPRF Architecture

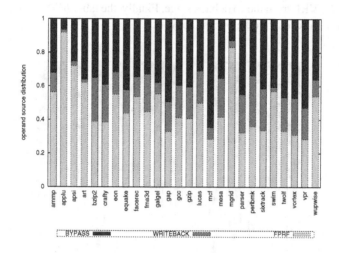

Fig. 3. Source of Integer Operands

operand read, execute, writeback and commit. To support FPRF access in the front-end, two stages are added to the pipeline: arbitration and FPRF Read. In the back-end the operand read stage disappears. This reduces the length to a single additional stage.

In the first new stage the source registers are analyzed to check for bank conflicts in the FPRF access. Conflicts stall all prior stages.

During rename, it is checked if the source registers have computed values. This is implemented via a bit vector with as many entries as logical registers. In the case of the Alpha ISA, modeled here, this requires maintaining two 32-bit vectors, one for the integer and another for the floating point registers. Each entry of this bit-vector indicates whether the corresponding logical register has a computed value. In case the computed value is available a read to the corresponding FPRF register is started.

During the arbitration cycle, priority is given to "older" instructions to access the operands. This makes sure that the front-end does not dispatch instructions to the instruction queues out-of-order.

The number of read ports for each bank has to be at least two, because some instructions must obtain both operands from the same FPRF bank.

Once arbitration has been performed the FPRF read occurs. After the instruction has read the available values it is inserted in the instruction queues. This happens during the *Queue* stage. At the same time, the register values are inserted into the VRF.

It is clear that the access rate to the FPRF is lower than to the centralized back-end physical register file. Lower access rate means that less conflicts will occur in the front-end and also that it will consume less energy. It was observed that the number of integer operands that are obtained from the FPRF is about 40% of the total while for floating point operands this number decreases to around 20%. Figures 3 and 4 show the distribution of integer and fp operand sources averaged over 100 million instructions for each Spec2000 benchmark. The boxes labeled *FPRF* account for operands that were read from the FPRF. The boxes labeled *WRITEBACK* are for operands that were written directly into the VRF from the writeback stage. Finally, the label *BYPASS* refers to those values that are sourced from the bypass network.

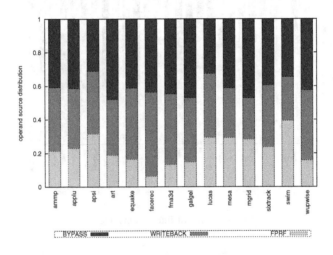

Fig. 4. Source of FP Operands

In the event of a branch misprediction the FPRF architecture behaves exactly like the MIPS R10000. First, the processor immediately aborts all instructions fetched along the mispredicted path. Next it restores the register mapping from the branch stack and finally, it starts fetching instructions from the correct path.

3.2   Read Sharing

Accesses to the same logical registers are often clustered during program execution. For example, many instructions use the same logical register for both register sources (eg ADD R2, R2, Rd). On the other hand it is also fairly common that the same register is sourced by several instructions without being written to. Such register accesses have a high probability of bank conflict. This suggests that conflicts in the access to the FPRF can be effectively reduced by using the technique known as read sharing [10]. Read sharing allows multiple reads of a same register to happen using a single local port which is connected to several global ports. Previous work on banked register files by Tseng et al. [21] has also used this approach. Read sharing will be evaluated later in the context of the FPRF architecture.

3.3 Writeback Filtering

In the FPRF implementation described so far all values are written back to the FPRF during the writeback stage. The front-end of the FPRF needs only to maintain copies for those physical registers that may be needed in the future. This includes all currently mapped registers and all registers that may be needed in the case of a misspeculation or exception recovery. The total number of registers that are generated by writeback is larger than these two numbers combined. Many registers that are renamed twice in a short interval do not appear in any of the mappings. Their storage will not be needed after the consumer reads the value. If this can be detected in time, then the writeback can be filtered and the write to the FPRF can be eliminated. To implement this strategy the processor needs to analyze the mapped registers in all rename checkpoints plus the current mapping and decide if the register that is being written back belongs to the processor state. Checkpoints need to be taken at all instructions that may cause a replay. There are many such instructions but the vast majority are conditional branches and loads. Registers that are not referenced anywhere are candidates to be filtered out during writeback.

One very interesting property of the writeback filtering concept is that lazy implementations can be built that are out of the critical path. The information whether an operand needs to be written back or not can be computed right after the rename stage. However, the operand itself will not be produced until the execution stage has completed, which is at least 5 cycles in the future. In general we can delay the computation of the filter mask a number of cycles equivalent to the distance between rename and writeback. However it has to be noted that delaying this computation will allow many unnecessary writebacks to happen because a physical register that is no longer necessary may appear as mapped in the filter mask even though it does not belong anymore to the current mapping.

This lazy writeback scheme allows the designer to propose slower but pipelined hardware structures to compute the filtering mask. For example, one proposal would be to use a slow multistage OR structure to compute the OR of the several checkpointed rename maps, assuming that a CAM-style renamer is being used. We expect such a structure to be slower, but also less power-hungry compared to precise proposals based on counters such as [13].

Section 5 presents performance results on writeback filtering for a scheme that computes the filter mask immediately. The lazy scheme has not been evaluated but it is expected to have similar performance.

4   Experimental Setup

For the evaluation of the FPRF architecture a heavily modified execution driven simulator based on SimpleScalar was used. The simulator executes binaries compiled for the Alpha ISA. Our benchmark suite consists of all benchmarks of the Spec2000 suite compiled with Digital *cc* using "-O2". We run the benchmark for 100 million of committed instructions.

First a baseline out-of-order microarchitecture with a centralized back–end RF was simulated and then extended using the enhancements proposed in this paper. Finally,

Table 1. Common parameters for all configurations

Fetch/Issue/Commit Width	4 instructions/cycle
Branch Predictor	Combined bimodal + 2-level
I-L1 size	32 KB, 4-way, 1 cycle latency
D-L1 size	32 KB, 4-way, 2 rd/wr ports, 2 cycle latency
D-L2 size	256 KB, 4-way, 2 rd/wr ports, 10 cycle latency
Memory Width / Latency	32 bytes / 100 cycles
Ports to the Register File	8 Read & 4 Write
Reorder Buffer Size	128
Integer/FP Physical Registers	160 / 160
Load/Store Queue	128 entries
Integer/FP Queue	32 entries / 32 entries
Integer/FP Functional Units	4 (latency 1) / 4 (latency 2)

we implemented the model described in [21] for comparison. The common parameters of all configurations are shown in Table 1.

To evaluate the proposal with the read sharing technique, the 6 configurations summarized in Table 2 and described below were simulated.

1. *Base-SHORT* is the optimal baseline configuration. Two baselines have been used to benchmark the proposal. Both Base-SHORT and Base-LONG simulate an out of order configuration with a full-ported centralized physical register file in the back-end. This architecture is based on the MIPS R10000 microarchitecture [1]. *Base-SHORT* simulates an architecture with a three-stage front-end and a six-stage back-end. This configuration has a pipeline that is one stage shorter than the FPRF pipeline. The performance will be higher not only because it does not have conflicts in accessing the register file, but also because the branch misprediction penalty is smaller than in the FPRF proposal. This is the reason whye Base-LONG is introduced.
2. *Base-LONG* is an architecture identical to Base-SHORT, but with an additional stage in the front-end. This model is identical to the FPRF model in number of cycles paid when a branch misprediction happens.
3. *FPRF-8B2R2W* is the FPRF with 8 banks, 2 read ports and 2 write ports each. It matches the configurations that Tseng et al. present in their paper on banked register files [21]. This configuration is the base case without optimizations. The pipeline depth is equal to Base-LONG.
4. *FPRF-8B2R2W-RS* is identical to the FPRF-8B2R2W configuration plus read sharing as described in Sect. 3.2.
5. *BMRF-OPT* is an optimistic implementation of the banking strategy described in [21]. This proposal needs to speculatively issue instructions to the functional units. If later a conflict occurs when the instruction wants to read its operands, a bubble is inserted in the pipeline for the conflicting instruction while the correct state of

Table 2. Main differences between configurations

Configuration	#Banks	Read Ports per Bank	Write Ports per Bank	Read Sharing	Bubble?	Pipeline Length
Base-SHORT	1	Unlimited	Unlimited	NO	-	9
Base-LONG	1	Unlimited	Unlimited	NO	-	10
FPRF-8B2R2W	8	2	2	NO	-	10
FPRF-8B2R2W-RS	8	2	2	YES	-	10
BMRF-OPT	8	2	2	YES	NO	10
BMRF-STALL	8	2	2	YES	YES	10

the microarchitecture is recovered. This model is optimistic because it assumes that there is no need to kill the full issue group and that the architecture can recover at once.

6. *BMRF-STALL* is the same implementation as *BMRF-OPT* but in this case it takes into account that a full bubble is inserted in the pipeline when a conflict occurs in the access to the register file and thus the full issue group is killed. This configuration approximates the work done in Tseng et al. [21] but, as will be verified later, it is still optimistic because some constraints have been left out of the simulated architecture.

For the evaluation of *writeback filtering* the FPRF-8B2R2W-RS architecture has been used.

5 Performance Evaluation

First, the amount of instruction level parallelism in the FPRF architecture is evaluated. On average, the architecture should be slower than both Base-SHORT and Base-LONG, because these architectures never stall and, in the case of Base-SHORT, the pipeline is shorter which makes branch recovery faster for this architecture.

The entire SPEC2000 benchmark suite was simulated for each of the 6 proposed configurations. To ease understanding, only averages and a few selected benchmarks are shown.

Figure 5 shows the IPC results relative to Base-SHORT. For the full SPEC average (rightmost column) the FPRF architecture, with no optimizations, is only 1.1% worse than Base-LONG. Applying the read sharing optimizations reduces this gap to 0.3%. Both differences are small. However, as can be seen, the behavior shows a lot of variance across benchmarks. For many applications the improvement is far from trivial.

The BMRF-OPT and BMRF-STALL implementations that model the banking strategy described in [21] performed on average 1.9% and 2.7% worse than Base-LONG, respectively. This is because stalls in the back-end have higher impact on performance than stalls in the front-end. These configurations use the read sharing optimization. Compared to our FPRF-8B2R2W-RS implementation, the BMRF implementations lose 1.6% and 2.3% IPC. Our values for the BMRF-STALL implementation are 1.5% better

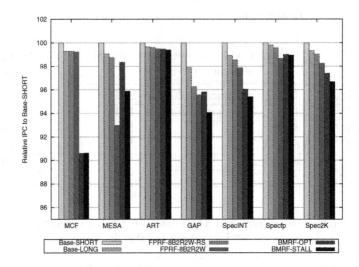

Fig. 5. Relative IPCs

than the values reported by the original paper. This variation is small and reasonable considering the differences between the modeled architectures: different numbers of physical registers and no operand partitioning between right and left register ports in the simulated BMRF models.

Comparing SpecINT and SpecFP results observes that there is much more IPC loss for integer benchmarks than for floating point programs. There are many reasons for this. For the specific case of Base-SHORT and Base-LONG, the high rate of mispredicted branches in integer applications is the cause for the IPC loss. FP programs have fewer branches and they are more predictable. Other columns of the plot show similar behavior. The reason FP programs have less IPC loss is because many more instructions are in the instruction queues and stalls in the front-end can be easily absorbed by the back-end. In addition , the fact that FP programs perform simultaneous FP and integer calculations helps to reduce the conflicts because both types of instructions access different register files.

Finally, the four selected benchmarks (MCF, MESA, ART and GAP) show how much the performance is dependent on program characteristics. The selected benchmarks show the widest variation observed across all of the SPEC benchmark.

5.1 Energy Consumption

One of the main benefits of implementing banking is to reduce the power consumption of the register file. The banking technique has long been known to reduce energy, but the complexity of control logic and potential loss in IPC, have precluded its use in register files. The banking scheme for the FPRF takes advantage of the reduced access rate to this structure when it is located in the front-end.

In this section the energy requirements of the register files of the FPRF architecture are evaluated. The energy model for the register file proposed by Rixner et al. [24] is used to obtain the energy of individual Rd/Wr access, which are then multiplied by the number of accesses to the register file. In the results, the values are averaged over all benchmarks of SPEC2000. Only the banked architectures are evaluated here using two different organizations for the instruction queues. The first type of back-end uses a centralized *Value Register File*. This is a heavily multiported structure (4 read ports and 10 write ports are needed). The results for this configuration are shown on the left side of Figure 6 relative to the power consumed by the centralized *Base-SHORT* architecture. The energy is given for both the FPRF and the VRF. The results show that the register distribution and banking techniques effectively reduce the energy consumed. Up to 94% of the Base-SHORT energy can be saved by combining the banked FPRF architecture with read sharing. On the other hand, although the VRF has fewer accesses and only 32 entries, its large number of ports increases the energy consumption compared to [21], which lacks a VRF. A detailed study of the VRF is outside of the scope of this paper.

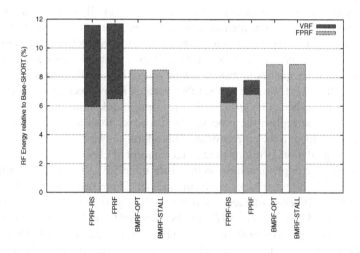

Fig. 6. Relative energy with a centralized VRF (left) and a distributed VRF (right)

An alternative organization based on multiple queues is also modeled, similar to the POWER4 microprocessor [25]. This microarchitecture uses multiple small instructions queues instead of one large centralized queue. Each queue has smaller issue capabilities (1-way) and less entries, which considerably reduces the number of ports (only 1 read port and 6 write ports, one for every source functional unit, are now required). 4 issue queues of 8 entries for both integer and FP datapaths are used here. The energy of such a distributed scheme is shown on the right side of Figure 6. Using this approach the energy of the VRF is reduced by 82% and the total energy is now 18% below the power consumed by the BMRF model. Using this IQ organization in the BMRF model would have no effect as, like the Power4, this model still requires the presence of a centralized register file that cannot be distributed. The new distributed scheme will have some IPC loss, but this will affect all architectures. A detailed analysis of this technique is outside the scope of this paper.

5.2 Writeback Filtering

The energy of the FPRF can be reduced even more by using *writeback filtering*. This technique has no effect on IPC, but it can have an impact on the energy because it reduces the number of writes into the register file. We measured the number of write-backs that can be filtered out from Spec2000 and observed that about 22% of all integer writebacks could be filtered out using this technique. The number of floating point writebacks is somewhat smaller, with 18% of all floating point results being short–lived values, which are candidates for writeback filtering. The energy reduction is expected to be somewhat smaller as only some writes but no reads are removed. For the full SPEC suite, the energy due to accesses to the integer register file is reduced by 12.1%. For SpecFP the energy due to accesses to the floating point register file is reduced by 13%.

These improvements are still limited. The main reason is the way load misses are handled. The need to replay load misses forces to add registers that appear in load rename maps to the rename stack where writeback filtering is analyzed. The number of loads is large and so is the number of additional maps. Not all implementations force loads to replay restarting from architectural state. Some processors use a scheme in which instructions that depend on a load are kept in the instruction queues until the load resolves in case they have to be reissued. This allows the processor to recover directly from the issue queues in case of a load scheduling conflict. In such a scheme less registers will be written back into the FPRF because many checkpoints will be gone. As an alternative, loads can be associated to previous checkpoints taken on branches. In the case of a load replay, the processor will recover from a previous checkpoint and re-execute the load in the correct order. This approach, which is expected to suffer marginal IPC degradation, is reminiscent of the concept of checkpointed commit [26].

In this alternative implementation on the FPRF up to 55% of all integer writebacks and up to 67% of all FP writebacks can be filtered out. These values are much closer to the expected gains. In terms of energy, this means that up to 30% of the energy can be saved in the integer FPRF and up to 47% can be removed from the floating point FPRF. These values may seem very large but it has to be remembered that the FPRF provides only 20-40% of the operands, so the bulk of accesses to this register file are writes. This is what makes the writeback filtering technique so attractive. On the downside, maintaining load-dependent instructions in the queue until the resolution of the load will have a negative impact on IPC because the queues will fill sooner. Therefore the checkpointing approach is preferable.

The use of *writeback filtering* is interesting because it allows to attack the energy problem from two perspectives. Positioning the physical register file in the front–end allowed to reduce the number of read accesses because many registers don't have computed values at this point. Writeback filtering, on the other hand, reduces the number of writeback accesses. Thus this combination of techniques allows simultaneous reduction in both read and write accesses.

6 Conclusions

This paper proposed and analyzed a new register file architecture combining the best of the Future File and the centralized physical register file. This combination reduces

RF access frequency and maintains fast misprediction recovery. It was shown that this architecture is well suited for the application of register file banking with read sharing and writeback filtering. Minimal IPC loss was observed when using a banked register file, considerably less compared to a previous proposal where the physical register file is in the back-end [21]. Only 0.3% of IPC was lost due to the banking conflicts, down from 1.6% in the previous proposal.

Writeback Filtering was proposed to reduce the number of writes to the register file. Two possible implementations of this were discussed. The first one used the architectural state to replay loads while the second replayed the loads by associating them with previous checkpointed branches.

The second configuration was able to remove about 60% of all writebacks, while the first only removed about 20%.

Acknowledgments

This work has been supported by the Ministry of Education of Spain under contract TIN–2004–07739–C02–01, the HiPEAC European Network of Excellence, the CEPBA and the Barcelona Supercomputing Center.

References

1. Yeager, K.C.: The MIPS R10000 superscalar microprocessor. IEEE MICRO **16** (1996) 28–41
2. Gonzalez, R., Cristal, A., Pericàs, M., Veidenbaum, A., Valero, M.: Scalable distributed register file. In: Proc. of the 4th Workshop on Complexity-Effective Design (WCED). (2004)
3. Levitan, D., Thomas, T., Tu, P.: The PowerPC 620 microprocessor: A high performance superscalar RISC microprocessor. In: Proc. of COMPCON'95. (1995) 285–291
4. Zyuban, V., Kogge, P.: The energy complexity of register files. In: Intl. Symp. on Low Energy Electronics and Design. (1998) 305–310
5. Seznec, A., Toullec, E., Rochecouste, O.: Register write specialization register read specialization: a path to complexity-effective wide-issue superscalar processors. In: Proc. of the 35th Intl. Symp. on Microarchitecture. (2002) 383–394
6. Park, I., Powell, M.D., Vijaykumar, T.: Reducing register ports for higher speed and lower energy. In: Proc. of the 35th Annual Intl. Symposium on Microarchitecture. (2002)
7. Kim, N.S., Mudge, T.: Reducing register ports using delayed write-back queues and operand pre-fetch. In: Proc. of the 17th ACM Intl. Conf. on Supercomputing. (2003)
8. Gonzalez, R., Cristal, A., Ortega, D., Veidenbaum, A., Valero, M.: A content aware integer register file organisation. In: Proc. of the 31th Intl. Symp. on Computer Architecture. (2004)
9. Cruz, J., González, A., Valero, M., Topham, N.: Multiple-banked register file architecture. In: Proc. of the 27th Intl. Symp. on Computer Architecture. (2000) 316–325
10. Balasubramonian, R., Dwarkas, S., Albonesi, D.: Reducing the complexity of the register file in dynamic superscalar processors. In: Proc of the 34th Intl. Symp. on Microarchitecture. (2001)
11. Zalamea, J., Llosa, J., Ayguadè, E., Valero, M.: Two-level hierarchical register file organization for VLIW processors. In: Proc of the 33th Intl. Symp. on Microarchitecture (MICRO-33). (2000) 137–146

12. Palacharla, S., Jouppi, N., Smith, J.: Complexity-effective superscalar processors. In: Proc. of the 24th Intl. Symp. on Computer Architecture. (1997)
13. Moudgill, M., Pingali, K., Vassiliadis, S.: Register renaming and dynamic speculation: an alternative approach. In: Proc. of the 26th. Intl. Symp. on Microarchitecture. (1993) 202–213
14. Balasubramonian, R., Dwarkadas, S., Albonesi, D.: Dynamically allocating processor resources between nearby and distant ilp. In: Proc. of the 28th Intl. Symp. on Computer Architecture. (2001) 26–37
15. Martinez, J., Renau, J., Huang, M., Prvulovic, M., Torrellas, J.: Cherry: Checkpointed early resource recycling in out-of-order microprocessors. In: Proc. of the 35th Intl. Symp. on Microarchitecture. (2002) 3–14
16. González, A., González, J., Valero, M.: Virtual-physical registers. In: Proc of the 4th Intl. Symp. on High-Performance Computer Architecture. (1998)
17. Lipasti, M.H., Mestan, B., Gunadi, E.: Physical register inlining. In: Proc. of the 31th Intl. Symp. on Computer Architecture. (2004)
18. Balakrishnan, S., Sohi, G.: Exploiting value locality in physical register files. In: Proc. of the 36th Intl. Symp. on Microarchitecture. (2003)
19. Smith, J.E., Pleszkun, A.R.: Implementation of precise interrupts in pipelined proccessors. Proc. of the 12th Intl. Symp. on Computer Architecture (1985) 34–44
20. Keltcher, C., McGrath, K., Ahmed, A., Conway, P.: The amd opteron processor for multiprocessor servers. IEEE MICRO 23 (2003) 66–76
21. Tseng, J., Asanovic, K.: Banked multiported register files for high-frequency superscalar microprocessors. In: Proc. of the 30th Annual Intl. Symp. on Computer Architecture. (2003)
22. Ponomarev, D., Kucuk, G., Ergin, O., Ghose, K.: Reducing datapath energy through the isolation of short-leved operands. In: Proc. of the 12th Intl. Conf on Parallel Architectures and Compiler Techniques. (2003)
23. Sami, M., Sciuto, D., Silvano, C., Zaccaria, V., Zafalon, R.: Exploiting data forwarding to reduce the power budget of vliw embedded processors. In: Proc. of the Conference on Design, automation and test in Europe. (2001) 252–257
24. Rixner, S., Dally, W.J., Khailany, B., Mattson, P.R., Kapasi, U.J., Owens, J.D.: Register organization for media processing. In: Proc. of the 6th Intl. Symp. on High Performance Computer Architecture. (2000) 375–386
25. Tendler, J., Dodson, S., Fields, S., H. Le, B.S.: Power4 system microarchitecture. IBM Journal of Research and Development 46 (2002)
26. Cristal, A., Valero, M., Gonzalez, A., LLosa, J.: Large virtual ROBs by processor checkpointing. Technical report (2002) Technical Report number UPC-DAC-2002-39.

Reducing Delay and Power Consumption of the Wakeup Logic Through Instruction Packing and Tag Memoization

Joseph Sharkey[1], Dmitry Ponomarev[1], Kanad Ghose[1], and Oguz Ergin[2]

[1] Department of Computer Science, State University of New York,
Binghamton, NY 13902-6000
{jsharke, dima, ghose}@cs.binghamton.edu
[2] Intel Barcelona Research Center,
Intel Labs, UPC, Barcelona, Spain
Oguzx.ergin@intel.com

Abstract. Dynamic instruction scheduling logic is one of the most critical components of modern superscalar microprocessors, both from the delay and power dissipation standpoints. The delay and energy requirement of driving the result tags across the associatively-addressed issue queue accounts for a significant percentage of the scheduler's overhead and also limits the design scalability. We propose two schemes to reduce the power consumption and the delays of the wakeup logic. Our first scheme – instruction packing – shares the associative part of an issue queue entry between two instructions, each with at most one non-ready source. As a result, the number of entries in the issue queue (and, hence, the length of the tag buses) can be reduced by a factor of two with almost no impact on the IPCs, because most instructions either enter the pipeline with at least one of their source operands ready, or do not make use of two source registers to begin with. Our second scheme – tag memoization – avoids driving the upper portion of the tags, if those bits did not change their values from what was driven on the same tag bus during the most recent broadcast. While instruction packing results in the reduced length of the tag buses, tag memoization reduced the number of tag lines that need to be driven. We evaluate our designs using detailed microarchitectural simulations of the SPEC 2000 benchmarks and the SPICE simulations of the issue queue layouts.

1 Introduction

Modern superscalar processors use out of order execution to exploit instruction level parallelism. The dynamic scheduling engine employed in such processors often uses associative logic embedded into the issue queue entries to wakeup instructions that are awaiting a result. This is accomplished by storing the address of the source registers within the issue queue entries and using the comparators that match the stored source register values against the address of the result that is driven on tag bus lines. A significant amount of energy dissipation results as the destination register address is driven on the tag busses. Energy dissipation occurs when the tag bus lines

B. Falsafi and T.N. Vijaykumar (Eds.): PACS 2004, LNCS 3471, pp. 15–29, 2005.
© Springer-Verlag Berlin Heidelberg 2005

are driven because of the charging and discharging of the wire capacitance of the tag line itself and the gate capacitance of the devices that implement the tag comparators. As wire capacitances dominate, a significant fraction of the energy spent in waking up instructions is attributed to the power used for driving the tag busses. This is particularly true if comparators that dissipate energy only on a match are used within the issue queue [27].

The scope of this paper is to propose two fairly orthogonal techniques for reducing the energy dissipated in driving the tag lines. Our first approach reduces the effective length of the tag bus lines and the number of comparator bits driven by essentially reducing the number of issue queue entries through the opportunistic packing of two instructions into a single issue queue entry. Our second approach avoids the power dissipated in driving the tag lines by not driving the higher order bits in the tag bus if their value matches the corresponding values last driven on the same tag bus. We validate the power savings achieved by using our techniques through the cycle-accurate simulations of SPEC 2000 benchmarks and the circuit simulations of the full-custom issue queue layouts.

2   Instruction Packing

In a traditional RISC-like processor where each instruction can have at most two register source operands, each issue queue (IQ) entry has two comparators, which allow the instruction to track the arrival of both sources by monitoring the tag buses. In general, however, such a design results is a grossly inefficient usage of the CAM logic, because of two reasons: 1) Many instructions have only one source register operand, and therefore do not require the use of two tags (and two comparators) in the first place, and 2) of the instructions with 2 source operands, a large percentage have at least one of the source operands ready at the time of dispatch, again rendering the second comparator unnecessary. Our simulations showed that on the average across SPEC 2000 benchmarks, about 83% of the dynamic instructions enter the scheduling window with at least one of their source operands ready.

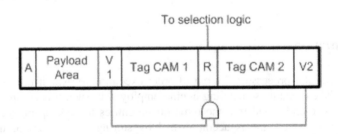

Fig. 1. Traditional IQ entry format

These statistics have been presented before [4] and researchers have proposed different solutions to optimize the IQ design based on this inefficiency. In [4], the non-uniform IQ entry formats were used, i.e. some entries have a full set of tag comparators, other entries have just one comparator, and yet other IQ entries have no comparators. In [28], the tag buses were subdivided into the slow buses and fast buses, such that the tag broadcast on the slow bus is delayed by one cycle. In this paper, we propose a different approach to optimizing the use of the CAM logic within the issue queue by packing multiple (two, for this paper) instructions into the same issue queue entry, effectively duplicating the RAM storage for these instructions (destination register addresses, literals, opcodes) and sharing the existing CAM logic. In effect, the aspect ratio of the issue queue changes: the number of issue queue entries become lower and the width of each entry goes up. In this section, we describe the details of our design.

Figure 1 shows a format of the issue queue entry used in traditional designs. The following fields comprise a single entry: a) entry allocated bit (A), b) payload area (opcode, FU type, destination register address, literals), c) tag of the first source, associated comparator (tag CAM word 1, hereafter just tag CAM 1, without the "word") and the source valid bit, d) tag of the second source, associated comparator (tag CAM 2) and source valid bit, and e) the ready bit. The ready bit, used to raise the request signal for the selection logic is set by AND-ing the valid bits of the two sources.

If at least one of the source operands is ready at the time of dispatch, the tag CAM associated with this instruction's IQ entry remains unused. To exploit this idle tag CAM, we propose to share one issue queue entry between two such instructions. An entry in the IQ can now hold one or two instructions, depending on the number of ready operands of the stored instructions at the time of dispatching. Specifically, if both source registers of an instruction are not available at the time of dispatch, the instruction is assigned an IQ entry of its own and makes use of both tag CAMs in the assigned entry to determine when its operands are ready. An instruction that has only one source register that is not available at the time of dispatch is assigned just one half of an IQ entry. The remaining half of the IQ entry may be used by another instruction that also has one of its source registers unavailable at the time of dispatch. Sharing an IQ entry between two instruction also requires the IQ entry to be widened to permit the payload parts of both instructions to be stored, along with the addition of flags that indicate whether the entry is shared between two instructions and the status of the stored instruction(s). Figure 2 shows the format of an issue queue entry that supports instruction packing. Each IQ entry is comprised of the "entry allocated" bit (A), the ready bit (R), the mode bit (MODE) and the two symmetrical halves: the left half and the right half. The structure of each half is identical, so we will use the left half for the subsequent explanations.

A left half of each IQ entry contains the following fields:

1. Left half allocated (AL) bit. This bit is set when the half-entry is allocated.
2. Source tag and associated comparator (Tag CAM). This is where the tag of the non-ready source operand for an instruction with at most one non-ready source is stored.
3. Source valid left bit (SVL). This bit signifies the validity of the source from part b), similar to traditional designs. This bit is also used to indicate if the instruction residing in a half-entry is ready for selection (as explained later)
4. Payload area. The payload area contains the same information as in the traditional design, namely: opcode, bits identifying the FU type, destination register address and literal bits. In addition, the payload area contains the tag of the second source.

Notice that the tag of the second source does not participate in the wakeup, because if an instruction is allocated to a half-entry, the second source must be valid at the time of dispatch. Compared to the traditional design, the payload area is increased by the number of bits used to represent a source tag.

The contents of the right half are similar. The ready bit (R) is used when an instruction with two non-ready source operands is allocated into the full IQ entry, as explained below. To summarize, each entry in the modified IQ is divided into a left half and a right half, each is capable of storing an instruction with at most one non-ready source operand, or the two halves can be used in concert to house an instruction with 2 non-ready source operands. In general, the issue queue entry can be in one of the following three states: 1) the entry holds a single instruction, both source operands of which were not ready at the time of dispatch, 2) the entry holds two (or one with another half free) instructions, each of which had at least one source operand ready at the time of dispatch, or 3) the entry is free. The "mode" bit, stored within each IQ entry as shown in Figure 2, identifies the state of the entry. If the mode bit is set to 1, then the entry maintains a 2-operand instruction, otherwise it either maintains one or two single-operand instructions or it is free.

Since each entry can hold up to two instructions, fewer IQ entries are needed. However, despite the fact that each entry in the modified IQ shown in Figure 2 is somewhat wider than the traditional queue entry (due to the replication of the Payload area and three extra bits – AL, AR, and MODE), the amount of CAM logic per-entry does not change. Each entry still uses only two comparators – those are either used by one instruction, which occupies full entry, or are shared by two instructions, each located in half-entry. In the next few subsections, we describe the details of this technique.

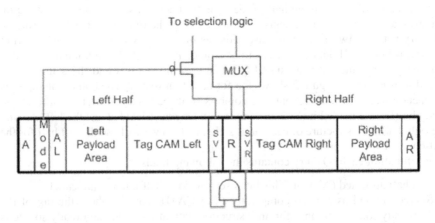

Fig. 2. Wakeup and Selection Logic Modified to Support Instruction Packing

2.1 Entry Allocation

To set up an issue queue entry for an instruction, the entry allocated bits corresponding to both halves (AL and AR), as well as the global "entry_allocated" bit

(A) are associatively searched in parallel with register renaming and checking the status of source physical registers. If the instruction is determined to have at most one non-ready source operand, the lowest numbered issue queue entry with at least one available half is allocated. If both halves are available within the chosen entry, then the instruction is written into the right half. After the appropriate half is chosen, both the "entry_allocated" bit of this half and the global A bits are reset. If an instruction is determined to have 2 non-ready source operands, then a full-sized entry is allocated, as dictated by the state of the A bits. The search for a full-sized and a half-sized entry occurs simultaneously, and the entry to be allocated is then chosen based on the number of non-ready source operands. This IQ entry allocation process is somewhat more complicated than similar allocation used in traditional designs, where just the A bits are associatively searched. However, there is no extra delay involved, because the searches occur in parallel. Similar issues with allocating the IQ entries are also inherent in other designs which aim to reduce the amount of associative logic in the queue by placing the instructions into the issue queue entries judiciously, based on the number of non-ready operands at the time of dispatch [4]. We will discuss what kind of information is written into the IQ for the various instruction categories later in the paper. But first, we describe how wakeup and selection are implemented in this scheme.

2.2 Instruction Wakeup

The process of instruction wakeup remains exactly the same as in traditional design for an instruction that occupies a full IQ entry (i.e. comes with 2 non-ready sources). Here, the ready bit (R) is set by AND-ing the valid bits of both sources. For instructions which occupy half of an IQ entry, the wakeup simply amounts to setting of the valid bit corresponding to the source that was non-ready when the instruction entered the IQ. The contents of the source valid bits are then directly used to indicate that the instruction is ready for selection (the validity of the second source is implicit in this case). The selection logic details are described next.

2.3 Instruction Selection

The process of instruction selection needs to be slightly modified to support instruction packing. To make the explanation easier, we assume that a 32-entry IQ is packed into a 16-entry structure, such that each entry is capable of holding two instructions with at most one non-ready source each, or one instruction with two non-ready sources. In a 32-entry IQ design, there are 32 request lines that can be raised by the awakened instructions – one line per IQ entry. In the instruction packing scheme, each of the two halves of each of the 16 entries requires a request line, thus retaining the same total number of request lines (32) and resulting in a similar complexity of the selection logic. In addition, the ready bits, used by the instructions allocated to full entries, also require request lines. Consequently, a straightforward implementation of the selection logic would require 48 (3x16) request lines, thus increasing the complexity, delay and power requirements of the select mechanism.

Such an undesirable elevation in the complexity of the selection logic can be avoided by sharing one request line between the R and the SVR bits. The shared request line is raised if at least one of the bits (the R or the SVR) is set. The R and the SVR bits are both connected to the shared request line through a multiplexor, which is controlled by the "mode" bit of the IQ entry (Figure 2). Consequently, the overall delay of the selection logic increases only slightly – by the delay of a multiplexor. Notice also that the MUX control signal (the "mode" bit) is available in the beginning of the cycle when the selection process takes place (the "mode" bit is set when the issue queue entry is allocated). The request line driven by the SVL bit is controlled by the p-device, whose gate is connected to the "mode" bit. This request line will be asserted only if the "mode" bit is set to 0 (indicating that the IQ entry is shared between two instructions) and the SVL bit is set to 1.

Note that the only part of the selection logic that is modified is the process of asserting the request lines. The rest of the selection logic is unchanged compared to the traditional designs. The overall delay of the selection logic is thus increased by the delay of the multiplexor, whose control signal is preset (as the value of the "mode" bit is available as soon as the IQ entry is allocated).

2.4 Instruction Issue

We define instruction issue as a process of reading the source operand tags of the selected instructions and starting the register file access (effectively moving the instruction out of the IQ). When a grant signal comes back corresponding to the request line, which was shared between the R and the SVR, the issue logic has to know which physical registers have to be read. Conventionally, this information is conveyed by the contents of the tag fields. However, the register tags of an instruction with two non-ready sources (i.e. the instruction that occupies full IQ entry) and the register tags of an instruction with one non-ready source are generally stored in different locations within the IQ entry. In the former case, the tags are stored in the tag fields connected to both comparators – one tag is stored in the left half of the entry and the other tag is stored in the right half of the entry. In the latter case, both tags are stored in the right half of the entry, such that the tag of the non-ready operand is connected to the comparator and the other tag is simply stored in the payload area. Given this disparate locations of the source register tags, how would the issue logic know which tags to use when the grant signal corresponding to a shared request line comes back?

One solution is, again, to use the contents of the "mode" bit and a few multiplexors. This will, however, slightly increase the delay of the issue / register access cycle. A better solution, which avoids the additional delays in instruction issuing altogether, is as follows. When an instruction with two non-ready sources is allocated to the issue queue, the tag, which is connected to the left half comparator, is also replicated in the payload area storage for the second tag in the right half. As a result, both tags will be present in the right half of the queue, so these tags can be simply used for register file access, without regard for the IQ entry mode.

2.5 Benefits of Instruction Packing

Instruction packing, as described in this section, has several benefits over the traditional issue queue designs in terms of layout area, access delays and power consumption.

The area of the issue queue decreases, because compared to the traditional designs, the amount of RAM storage does not change (we use twice as fewer entries, but each entry has about twice the amount of RAM), but the amount of associative logic is reduced by a factor of two.

The delay of the wakeup logic is reduced, because the tag buses become much shorter and the capacitive loading on these buses is also significantly reduced – the delay in driving the tag bus (which is a major component of the wakeup latency) is roughly reduced by half. Furthermore, shorter bitlines can potentially reduce the IQ access delays during instruction dispatching (setting up the entries) and issuing (reading out the register tags and literals). Finally, for the same reasons the power consumption is also reduced. Another potential reason for the reduction in the power consumption has to do with the use of fewer comparators. In the instruction packing, the tags of the source registers ready during dispatching are never associated with the comparators. In the traditional designs, each and every source tag is hooked up to a comparator. Unless these comparators are precharged selectively (based on whether or not a given IQ slot is awaiting for the result), unnecessary dissipations can occur than comparators associated with the already valid sources continue to fire.

In the result section, we quantify these savings using detailed simulations of SPEC 2000 benchmarks and also circuit simulations of the IQ layouts. Notice that all these benefits are achieved with essentially no degradation in the IPCs (committed Instructions Per Cycle). This is because most instructions (our results show 83%) have at least one of their sources ready at the time of dispatch, thus rendering the performance loss due to the smaller number of IQ entries negligible.

3   Tag Memoization

The tag memoization scheme exploits the fact that the higher-order bits of the tags that are broadcasted within a short duration of each other are likely to be the same. The idea here is to conserve power expended in driving the tag by not driving the higher-order tag bits if they happen to match the higher-order tag bits that were driven on the same bus during the previous broadcast. The tag comparator used to match the tag on the bus is broken into two separate comparators, say Cu and Cl, to match the higher-order bits and the remaining lower-order bits, respectively. A 1-bit latch is inserted in between to remember if there was a match in the higher order bits with the previous broadcast. The match signal for an entry is derived by AND-ing the output of this latch with the output of the comparator for the lower order bits. Figure 3 depicts this logic. We now describe this scheme in some detail.

Let Lb designate the latch used within an IQ entry to remember the match with the upper order bits driven on bus b with the tag value stored within a register operand field of the IQ entry. The tag driver logic for tag bus b also uses a latch array, Ub, to remember the upper order bits of the tag pattern that was driven onto the tag bus b. The following two cases arise when a tag value is to be driven on a tag bus:

If the upper bits driven on the bus b in the next broadcast match Ub, then only the lower order bits are driven on the tag bus. Entries that match the lower order bits and have their latch Lb set now produce a match signal. If, however, the upper order bits driven on the bus b do not match the contents of Ub, then the following actions are taken concurrently:

- The reset line shown in Figure 3 is driven to clear the contents of latch Lb in all of the IQ entries.
- Both upper and lower order bits of the tag are driven out on bus b.
- Ub is updated.

Clearing Lb in this case allows each entry to produce a match based on all of the tag bits - both upper order and lower-order bits.

The tag memoization scheme saves power by not driving the upper order bits of a tag bus whenever possible. The power savings are somewhat defeated by the need to drive the reset line on each tag bus, by the need to maintain the Ub latch, and dissipations within the Lb and the AND-gate used within each entry. One can save additional power dissipation by using the contents of Lb to disable Cu once Lb is set. Doing so prevents Cu from dissipating any power from false matches with the values floating on the upper order bits of the bus.

From a delay standpoint, the AND-ing of Lb with the output of Cl adds a slight delay in the generation of a request signal from matching entries. This added delay is however compensated to some extent by the smaller delay of Cl. (Cl has a smaller response time compared to that of a comparator that compares all bits of the tag value.)

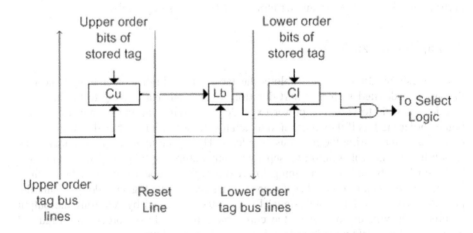

Fig. 3. Tag comparator configuration for the tag memoization scheme

One can force additional savings from the memoization scheme by assigning tag broadcasts to a bus whose Ub matches the upper order bits of the tag value to be driven. We call this "intelligent" tag bus assignment. There is, of course, some energy overhead in assigning tag broadcasts to specific buses in this case. Another possibility, and one that we have not explored here, is to assign the tag values sequentially to instructions. This is possible in datapaths that use the ROB slots as physical registers or have rename buffers that are assigned from a circular FIFO. In this paper, we only considered a datapath with the unified physical/architectural register file, where the allocation of physical registers can occur from anywhere within the register file, as dictated by the free list.

The approach just described can be generalized to accommodate the segmentation of the tag comparator into more than two parts requiring an intervening latch in between consecutive segments. For example, an 7 bit tag comparator can be segmented into three parts: Ca(upper order two bits), Cb(next two bits), and Cr (remaining 3 bits). This arrangement requires two latches: one between Ca and Cb and another between Cb and Cr; there is a reset line for each latch. These latches may be set independently, allowing for the gating off of either set of bits, or both. The match signal is derived by AND-ing the contents of the intervening latches and the output of the comparator segment covering the lower order bits.

Table 1. Configuration of the simulated processor

Parameter	Configuration
Machine width	4-wide fetch, 4-wide issue, 4 wide commit
Window size	Issue queue: as specified,48 entry load/store queue, 96–entry ROB
Function Units and Latency (total/issue)	4 Int Add (1/1), 2 Int Mult (3/1) / Div (20/19), 2 Load/Store (2/1), 2 FP Add (2), 2 FP Mult (4/1) / Div (12/12) / Sqrt (24/24)
Physical Registers	128 combined integer + floating-point physical registers
L1 I–cache	64 KB, 1–way set–associative, 128 byte line, 1 cycles hit time
L1 D–cache	64 KB, 4–way set–associative, 64 byte line, 2 cycles hit time
L2 Cache unified	2 MB, 8–way set–associative, 128 byte line, 6 cycles hit time
BTB	2048 entry, 2–way set–associative
Branch Predictor	Combined with 1K entry Gshare, 10 bit global history, 4K entry bimodal, 1K selector
Branch Mispred. Penalty	8 cycles minimum
Memory	128 bit wide, 150 cycles first chunk, 1 cycles interchunk
TLB	32 entry (I), 128 entry (D), fully associative, 12 cycles miss latency

4 Simulation Methodology

Our simulation environment includes a detailed cycle accurate simulator of the microarchitecture and cache hierarchy. While our simulator was developed from scratch, it uses the same binaries, system call interface and tools as the MIPS-like Simplescalar PISA ISA. All benchmarks were compiled with gcc 2.6.3 (compiler options: -O2) and linked with glibc 1.09, compiled with the same options. All simulations were run on a subset of the SPEC 2000 benchmarks consisting of 8 integer and 7 floating-point benchmarks. In all cases, predictors and caches were warmed up for 1 billion committed instructions and statistics were gathered for the next 200 million instructions. Table 1 presents the configuration of the baseline processor.

For estimating the delay, energy and area requirements, we deigned the actual VLSI layouts of the issue queue and simulated them using SPICE. The layouts were designed in a 0.18 micron 6 metal layer CMOS process (TSMC) using Cadence design tools. A Vdd of 1.8 volts was assumed for all the measurements.

5 Results

5.1 Instruction Packing

Table 2 shows the IPC loss due to instruction packing. These results are displayed in the form of a table rather than a graph because IPC differences are too small to be noticeable on the traditional bar graph. The columns, in order, show IPC results with a 32-entry issue queue, a 16-entry issue queue with instruction packing, an 8-entry IQ, and a 4-entry IQ with packing. The results show that a 16-entry issue queue utilizing instruction packing performs within 0.5% of a traditional 32-entry IQ. The configuration with an 8-entry queue packed into 4 wider entries is only shown to demonstrate that packing does not significantly degrade the performance even for very small issue queues. Here, for example, the performance loss is only 5.3% on the average.

Table 2. IPC for 32 and 8-entry traditional queues as compared to 16 and 4-entry queues supporting instruction packing

Benchmarks	32IQ	16IQ_PACK	8IQ	4IQ_PACK
Gzip	1.544	1.587	1.594	1.412
Vpr	1.463	1.447	1.279	1.125
Gcc	1.128	1.128	1.105	1.032
Mcf	0.444	0.442	0.398	0.343
Parser	1.317	1.304	1.234	1.118
Vortex	1.996	2.001	1.908	1.672
bzip2	1.594	1.546	1.480	1.272
Twolf	1.209	1.161	1.104	0.968
Wupwise	2.212	2.212	1.924	2.212
Swim	1.511	1.511	1.412	1.269
Mgrid	1.218	1.218	1.210	1.218
Applu	1.337	1.337	1.345	1.278
Mesa	1.786	1.786	1.580	1.786
Art	0.399	0.400	0.332	0.243
Equake	1.724	1.723	1.522	1.312
IntAvg	**1.337**	**1.327**	**1.263**	**1.118**
FPAvg	**1.455**	**1.455**	**1.332**	**1.331**
Average	**1.368**	**1.363**	**1.279**	**1.211**

Fig. 4. Layout of a CAM bitcell (left) vs. an SRAM bitcell (right)

CMOS layouts of both the 32-entry traditional queue and the 16-entry packing queue show a 26.7% reduction in the issue queue area due to the use of instruction packing. Packing effectively reduces the number of CAM bitcells by half, while increasing the number of SRAM bitcells in each row (but leaving the total number of SRAM bitcells in the IQ practically unchanged). Figure 4 presents the layouts of a CAM bitcell (left) and an SRAM bitcell (right).

As presented in Table 3, instruction packing achieves a 21.6% reduction in the wakeup delay (when a 32-entry IQ is packed into a 16-entry IQ). This delay reduction comes mainly from the shorter and lower-capacitance tag busses.

Table 3. Delays of a 16-entry queue supporting instruction packing compared to a 32-entry traditional queue

	Tag-Bus Drive (ps)	Comparator Output (ps)	Final Match Signal (ps)	Total Delay (ps)
32-entry	224	219	126	569
16-entry Packing	131	201	114	446
Savings:	41.5%	8.2%	9.5%	**21.6%**

Finally, the instruction packing saves energy due to the presence of half as many tag comparators and shorter tag-busses. SPICE simulations show the 16-entry packing queue saves 37.99% of total wakeup power as compared to a traditional 32-entry queue, most of it coming from the savings in the tag bus drive energy (we present the detailed per-benchmark results in Figure 6, Section 5.3).

5.2 Tag Memoization

Tag memoization does not impact IPC because it does not interrupt, hinder, or change the order of tag broadcasts in any way. The power savings of tag memoization comes from its ability to match the most significant bits of the tags on each bus from one broadcast to the next. Thus, it is important to consider how often these tag bits match. Figure 5 presents the number of most significant bits (MSBs), for each tag broadcast,

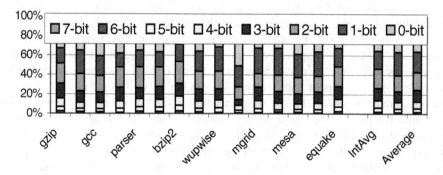

Fig. 5. Number of most significant bits matching those of the previous tag broadcast on each tag bus

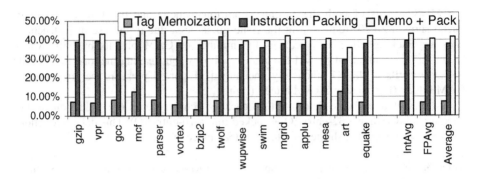

Fig. 6. Wakeup power savings

that match those of the previous tag broadcast on that bus. Since two bits match 43% of the time, we consider 2-bit matches for the remainder of this discussion.

On a configuration with two separate intermediate latches, the two MSBs match an average of 48.1% of the time, while the next two MSBs match an average of 24.5% of the time. Accounting for the extra line that must be driven every time one of these latches must be reset, the total power savings from such a variation of tag memoization is 7.3% of total tag-broadcast power. This power savings can be improved by selectively arbitrating for tag busses so as to maximize tag matches. Such an "intelligent" tag bus assignment is able to achieve tag broadcast power savings by as much as 12.8%.

5.3   Combining Instruction Packing with Tag Memoization

Instruction packing and tag memoization are two orthogonal approaches to reducing the power consumption of instruction wakeup. Instruction packing aims to reduce the length of the tag-broadcast while tag memoization aims to reduce the number of tag busses driven. The 16-entry queue that supports instruction packing and uses 2 + 2 tag memoization (where we segment the 7-bit comparator into 3 segments – two higher order segments, each 2 bits wide, hence 2+2, and a 3-bit segment for the lower order bits) with judicious bus arbitration (where the bus for a tag broadcast is selected to maximize the likelihood of matches in the most significant bits) reduces the wakeup power by 44.74% as compared to a traditional 32-entry queue. If the random bus selection is used, then the power reduction is about 41%. The per-benchmark results, showing power savings of both schemes in isolation, as well as the combined power savings, are presented in Figure 6. In this figure, the tag memoization results are presented for the scheme where the tag buses are randomly allocated. As seen from Figure 6, the combined power savings from these two schemes is not additive. This is simply because instruction packing changes the relative position of a source register address over the tag busses, changing the pattern of tag matching, thus impacting the effectiveness of tag memoizatoin. (Higher power savings (not shown) are achieved in the combined scheme when intelligent tag bus assignment is used.)

6 Related Work

Researchers have proposed several ways to reduce the power consumption of the issue logic. Dynamic adaptation techniques [22,23,24,25] partition the queue into multiple segments and deactivate some segments periodically, when the applications do not require the full issue queue to sustain the commit IPCs. Energy-efficient comparators, which dissipate energy predominantly on a tag match were proposed in [26,27]. Also in [26], the issue queue power was reduced by using zero-byte encoding and bitline segmentation. In [29], the associative broadcast is replaced with indexing to only enable a single instruction to wakeup. This exploits the observation that many instructions have only one consumer.

The observation that many instructions are dispatched with at least one of their source operands ready is not new – it was used in [4], where the scheduler design with reduced number of comparators was proposed. In that scheme, some IQ entries have two comparators, others have just one comparator, and yet others have zero comparators. Despite significant reduction in the number of comparators, the size of the issue queue, and thus the length of the tag busses, was not reduced. In addition, the last-tag speculation mechanism introduced in [4] requires the extra logic to handle possible mispredictions. In [28], the tag buses were categorized into fast buses and slow buses, such that the tag broadcast on the slow bus takes one additional cycle. The design again relied on the last-arriving operand prediction to hook the last arriving operand (which actually identifies when the instruction wakes up) to the fast bus to avoid the wakeup delays.

One approach to reducing scheduling complexity involves pipelining the scheduling logic into separate wakeup and select cycles [2,8]. It is shown in both [2] and [8] that naively pipelining the scheduling logic doesn't provide for the back-to-back execution of dependent instructions and thus significantly degrades performance. To overcome this, [2] uses the status of an instruction's grandparents to wakeup the instruction earlier in a speculative manner. Kim and Lipasti [8] proposed grouping of two (or more) dependent single-cycle operations into so-called Macro-OP (MOP), which represents an atomic scheduling entity with multi-cycle execution latency. A smaller issue queue can be used in this design, because the instructions forming the Macro-OP share the same issue queue entry. The concept of dataflow mini-graphs [21] is similar to Macro-Op scheduling in that groups of instructions are scheduled together. The order of instructions within the mini-graph are determined statically and the scheduler only considers "handles", or groups of instructions, for scheduling. This relies on re-compilation of code to generate these "handles" in the binary.

Other proposals have introduced new scheduling techniques with the goal of designing scalable dynamic schedulers to support a very large number of in-flight instructions [5, 6, 9, 14, 20]. Brown et.al. [7] proposed to remove the selection logic from the critical path by exploiting the fact that the number of ready instructions in a given cycle is typically smaller than the processor's issue width.

Scheduling techniques based on predicting the issue cycle of an instruction [10, 11,12,13,15,16,18] remove the wakeup delay from the critical path and remove the CAM logic from instruction wakeup, but need to keep track of the cycle when each physical register will become ready. In [17], the wakeup time prediction occurs in parallel with the instruction fetching.

7 Concluding Remarks

We proposed two orthogonal schemes to reduce the power consumption of the wakeup logic. Instruction packing combines two instructions within the same issue queue entry if both instructions have at most one non-ready source operand at the time of dispatch. Consequently, the number of issue queue entries, and thus the length of and the capacitive loading on the tag busses, can be reduced substantially, leading to faster access and lower power dissipation. In addition, the layout area of the issue queue is also reduced. Tag memoization avoids driving the portion of the tag if it did not change from what was previously driven on the same tag bus. Combined, the two techniques result in about 45% reduction in the wakeup power. Additionally, instruction packing also achieves 26% reduction in the issue queue layout area and 21% reduction in the wakeup delay. The delay of the selection logic increases only slightly - by the delay of a single multiplexer (with the pre-set control signal). Thus, significant overall reduction in the scheduler delay, and thus higher frequency, can be also realized.

References

[1] S. Palacharla, et. al., "Complexity-Effective Superscalar Processors", in the Proc. of the Int'l Symp. on Computer Architecture, 1997

[2] J. Stark, et. al., "On Pipelining Dynamic Instruction Scheduling Logic", in the Proc. of the Int'l Symp. on Microarchitecture, 2000

[3] Burger, D. and Austin, T. M., "The SimpleScalar tool set: Version 2.0", Tech. Report, Dept. of CS, Univ. of Wisconsin-Madison, June 1997 and documentation for all Simplescalar releases.

[4] D. Ernst, T. Austin, "Efficient Dynamic Scheduling Through Tag Elimination", in the Proc. of the Int'l Symp. on Computer Architecture, 2002.

[5] E. Brekelbaum et. al., "Hierarchical Scheduling Windows", in the Proc. of the Int'l Symp. on Microarchitecture 2002.

[6] A. Lebeck et. al. A Large, "Fast Instruction Window for Tolerating Cache Misses", in the Proc. of the Int'l Symp. on Computer Architecture, 2002.

[7] M. Brown, J. Stark, Y. Patt. "Select-Free Instruction Scheduling Logic", in the Proc. of the Int'l Symp. on Microarchitecture 2001.

[8] I. Kim and M. Lipasti, "Macro-Op Scheduling: Relaxing Scheduling Loop Constraints", in the Proc. of the Int'l Symp. on Microarchitecture 2003.

[9] A. Cristal, et.al., "Out-of-Order Commit Processors", in the Proc. of the Int'l Symp. on High Performance Computer Architecture, 2004.

[10] D. Ernst, A. Hamel, T.Austin, "Cyclone: a Broadcast-free Dynamic Instruction Scheduler with Selective Replay", in the Proc. of the Int'l Symp. on Computer Architecture'03

[11] Hu, J., Vijaykrishnan, N., Irwin, M., "Exploring Wakeup-Free Instruction Scheduling", in the Proc. of the Int'l Symp. on High Performance Computer Architecture, 2004.

[12] R.Canal, A. Gonzalez, "A Low-Complexity Issue Logic", in the Proc. of the Int'l Conference on Supercomputing, 2000.

[13] R.Canal, A.Gonzalez, "Reducing the Complexity of the Issue Logic", in the Proc. of the Int'l Conference on Supercomputing 2001.

[14] S. Raasch, N.Binkert, S.Reinhardt, "A Scalable Instruction Queue Design Using Dependence Chains", in the Proc. of the Int'l Symp. on Computer Architecture, 2002.

[15] J. Abella, A.Gonzalez, "Low-Complexity Distributed Issue Queue", in the Proc. of the Int'l Symp. on High Performance Computer Architecture, 2004.

[16] P. Michaud, et.al. "Data-Flow Prescheduling for Large Instruction Windows in Out-of-Order Processors", in the Proc. of the Int'l Symp. on High Performance Computer Architecture, 2001.

[17] T. Ehrhart, S. Patel, "Reducing the Scheduling Critical Cycle using Wakeup Prediction", in the Proc. of the Int'l Symp. on High Performance Computer Architecture, 2004.

[18] Y. Liu, et. al., "Scaling the Issue Window with Look-Ahead Latency Prediction", in the Proc. of the Int'l Conference on Supercomputing 2004.

[19] Z. Chishti, T. Vijaykumar, "Wire Delay Is Not a Problem for SMT", in the Proc. of the Int'l Symp. on Computer Architecture 2004.

[20] S. Srinivasan et. al. "Continual Flow Pipelines", in the Proc. of the Int'l Conference on Architectural Support for Programming Languages and Operating Systems, 2004.

[21] A. Bracy, et. al. "Dataflow Mini-Graphs: Amplifying Superscalar Capacity and Bandwidth", in the Proc. of the Int'l Symp. on Microarchitecture 2004.

[22] A. Buyuktosunoglu, et.al.,"A Circuit-Level Implementation of an Adaptive Issue Queue for Power-Aware Microprocessors", GLSVLSI, 2001.

[23] D.Folegnani, A.Gonzalez, "Energy-Effective Issue Logic", in the Proc. of the Int'l Symp. on Computer Architecture, 2001.

[24] D.Ponomarev, G.Kucuk, K.Ghose, "Reducing Power Requirements of Instruction Scheduling Through Dynamic Allocation of Multiple Datapath Resources", in the Proc. of the Int'l Symp. on Microarchitecture 2001.

[25] A. Buyuktosunoglu et.al., "Energy-Efficient Co-adaptive Instruction Fetch and Issue", in the Proc. of the Int'l Symp. on Computer Architecture, 2003.

[26] D.Ponomarev, et.al., "Energy-Efficient Issue Queue Design", in IEEE Transactions on VLSI Systems, November 2003.

[27] D.Ponomarev, et.al., "Energy-Efficient Comparators for Superscalar Datapaths", IEEE Transactions on Computers, July 2004.

[28] I.Kim, M.Lipasti, "Half-Price Architecture", in the Proc. of the Int'l Symp. on Computer Architecture, 2003.

[29] M.Huang et.al., "Energy-Efficient Hybrid Wakeup Logic", in the Proc. of the Int'l Symp. on Low-Power Electronics and Design, 2002.

Bit-Sliced Datapath for Energy-Efficient High Performance Microprocessors

Sumeet Kumar, Prateek Pujara, and Aneesh Aggarwal

ECE Department, Binghamton University,
Binghamton, NY 13902
{skumar1, ppujara1, aneesh}@binghamton.edu

Abstract. In the recent years, both power and performance have become important in the design of microprocessors. In this paper, we investigate exploiting the small-sized data values for energy-efficient high performance microprocessors. For this purpose, we bit-slice the execution core (which includes the functional units, register files, data caches, and forwarding logic), so that small portions of the data are operated upon in different bit-slices. The bit-slices operating upon the higher order bits are activated only if required, saving significant energy consumption. We also investigate further advantages facilitated by bit-slicing such as energy savings obtained by reducing the number of ports provided in the higher order bit-slices and by "shutting off" bit-slices to reduce leakage energy consumption. We found that a significant energy saving can be obtained in the register file (about 20%) and the Level-1 Data Cache (about 30%) with a negligible loss of only about 2% in the instruction throughput. Our studies also showed almost 20% savings in the register file leakage energy consumption, when the unwanted higher order bit-slices are "turned off". Bit-slicing also helps in reducing the latency of the different hardware structures, which can facilitate clock speed improvements.

1 Introduction

Energy consumption has emerged as an important criteria in the design of microprocessors [8]. The major contributor to the overall energy consumption in a chip is the dynamic energy consumption. However, the leakage energy consumption is also on the rise [6], and has started to become a concern in the microprocessor designs. Dynamic energy consumption results from the activity in a processor and is caused by the charging and discharging of the capacitive loads in the processor. Leakage energy consumption, on the other hand, is the result of shrinking transistor sizes which leads to increased sub-threshold current.

One important approach towards reducing the energy consumption in a processor, while not hurting the performance, is to limit the amount of unnecessary work performed by the processor. Clock gating [2] is a good example of this approach, where the hardware that does not need to be activated during an operation is not provided with the clock signal. The architecture presented in

B. Falsafi and T.N. Vijaykumar (Eds.): PACS 2004, LNCS 3471, pp. 30–45, 2005.

this paper is in the same spirit, *i. e.* limit the amount of unnecessary work performed by the processor. For this, we exploit the small-sized data values. Studies [1][2][11] have shown that a significant number of data values of small size and have a large number of leading zeros. In this paper, we bit-slice the processor datapath. In the bit-sliced architecture, a particular bit-slice operates only on certain data bits, and other bit-slices operate on other data bits. In this architecture, operations in the higher end bit-slices are performed only if required, thus reducing the unnecessary work and saving energy. Even though there are many techniques in the literature that exploit the narrow-width data property to various effects, bit-sliced datapath has only been proposed to a limited extent in [4], and not to the extent to which we bit-slice the execution core datapath. Bit-slicing the datapath also has the potential of improving the clock speed by reducing the access latencies of each of the bit-sliced hardwares. We experiment with a 32-bit RISC architecture, however, the benefits of bit-slicing are expected to increase as the processors use wider data sizes (64-bit processors and beyond). To motivate the approach, we present the sizes of the data values in the processor.

1.1   Data-Sizes

We measure the operand sizes for the integer instructions operating on the integer data values. For the measurements, we separate the integer instructions into simple (such as Add, Subtract, etc.), complex (such as Multiply, Divide, and Shifts), and load/store instructions. Figures 1(i), 1(ii), and 1(iii) give the percentage of simple, complex, and load/store instructions, respectively, that have either one or both operands of size greater than 8, 16, and 24 bits for a 32-bit RISC architecture. The legend for these graphs is given in Figure 1(iii). Operands of sizes greater than 24 bits also include the negative values. Note that, in the figures, the instructions that have at least 1 operand less than or equal to 8 bits (in the second bar) can have the other operand of a larger size. For the load and store instructions, we also measure the size of the values loaded from and store into the data cache, shown in Figure 1(iv).

Figure 1(i) shows that there are about 60% of simple instructions that have at least 1 operand that is greater than 16 bits (*i.e.* about 40% of the simple instructions have both the operands less than or equal to 16 bits). When both the operands are considered, there are only about 10% of the simple instructions that have both the operands greater than 16 bits, suggesting that, an instruction rarely operates on the higher order bits (there may be a few cases with a carry generated by the lower bits). When considering the complex instructions, in Figure 1(ii), there are considerably fewer instructions (about 30% on an average) that have at least one operand greater than 16 bits in size. For the complex instructions as well, there are only about 10% of instructions that have both their operands of size greater than 16 bits. The load and store instructions are different and they always have one operand (among the operands used for effective address calculation) that requires almost 32 bits for representation. Considering the size of the values that are loaded from the data cache and stored into the data cache,

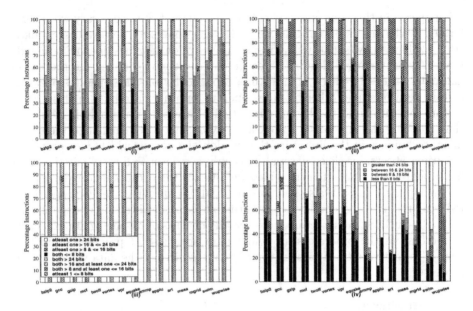

Fig. 1. Operand Sizes of Integer Instructions for (i) Simple; (ii) Complex; (iii) Load/Store Instructions (1st Bar is for One Operand and 2nd Bar is for Both the Operands); and (iv) Sizes of Loaded and Stored Values

there are only about 50% of the values that have a size that is greater than 16 bits (these include the negative values as well). The percentage of wider values loaded from the cache and stored into the cache is relatively higher for the FP benchmarks (`applu`, `art`, `ammp`, `mesa`, `mgrid`, `swim`, and `wupwise`), because they load and store floating point values that typically use the entire 32 (64 bits for double precision) for representation.

Figure 1 suggests that significantly small number of operations are performed on the upper bits of operands, and motivate a bit-sliced architecture, where the higher order bits are operated upon only when required.

1.2   Related Work

There has been some past work that uses the *data widths* of the values (generated in a program) to various effects. The SIMD paradigm takes advantage of the narrow width operands to improve the performance of the multimedia applications [10][13]. The Dynamic Zero Compression (DZC) cache [18] reduces cache energy by exploiting the small-width property of values stored in the data cache, by using a single bit to indicate that a full byte is zero.

Brooks [2] and Gabriel [11] use the small-width operand sizes to dynamically pack different small-width data values and perform simultaneous operations on them in order to improve performance, and reduce power consumption. Canal et. al. [4][5] proposed two approaches to exploit narrow-width operands. In [4], they considered a byte-serial (8-bit) or a semi-parallel (16-bit) pipeline to exploit the narrow-width data at the architectural level. The idea is to append extension bits to data residing in the caches and registers to reflect the significant part of the data, and only load, store, or compute on the useful bytes, thus reducing switching activities. However, the limited datapath width provided can lead to significant performance loss when processing operands of a larger bitwidth (32 bits or more). In [5], the authors rely on profiling information used along with static value range propagation analysis to discover useful range of operand-width, and re-encoding operands with narrower opcodes. Compiler-level efforts to exploit the narrow operand widths have been proposed in [12][16][14].

To the best of our knowledge, no one has yet investigated the performance of a bit-sliced execution core to exploit the narrow-width properties of operands to the extent that is being performed in this paper.

1.3 Contributions of This Work

The main focus of this paper is to study the performance of a bit-sliced datapath in terms of its impact on the instruction throughput and the energy savings for a high performance processor. Our studies showed that about 20% and 30% dynamic energy savings can be obtained in the register file and the data cache (by bit-slicing them), respectively, for a 2-way bit-sliced datapath. However, the reduction in the instruction throughput is only about 2%, compared to a non-bit-sliced datapath. We also investigate the reasons for IPC loss in a bit-sliced datapath and recover some of the lost IPC using performance enhancement techniques such as early resolution of branches. We also investigate how bit-slicing can facilitate further reduction in energy consumption. For this, we propose reducing the number of ports in the higher order bit-slices of the storage elements such as the register file and the data cache. To reduce the leakage energy consumption, we propose "shutting off" parts of higher order bit-slices that do not store significant data. With these energy reduction techniques, the overall dynamic energy consumption can be reduced by about 25% in the register files and about 32% in the data cache, and the register file leakage energy consumption can be reduced by about 20%.

The rest of the paper is organized as follows. Section 2 presents the bit-sliced execution core architecture and its impact on performance and energy consumption. Section 3 presents and discusses both the IPC and the energy consumption results. Section 4 proposes selective delays technique to recover some of the IPC loss incurred. Section 5 presents techniques to further reduce the dynamic and static energy consumption. We conclude in Section 6.

2 Bit-Sliced Execution Core Datapath

2.1 Basic Architecture

In a bit-sliced execution core, each wide integer ALU is partitioned into smaller width ALUs, the integer register file is partitioned into multiple smaller width register files, and even the data cache is partitioned into multiple smaller width data caches. For instance, for a 32-bit word machine, each 32-bit ALU can be partitioned into 2 16-bit ALUs, the integer register file can be partitioned into 2 banks, each of size 16 bits, and the data cache can be partitioned into 2 data caches, where each bank stores 16 bits of a word. We call each such partitioned hardware module as a *bit-slice*. In the bit-sliced datapath, the lowest bit-slice only operates on the lowest bits of the operands, and the next higher bit-slice operates on the next higher bits, and so on. A 2-way bit-sliced execution core datapath (for 2 ALUs and 1 data cache port) is shown in Figure 2, where the ALUs, the register file, the bypass network, and the L1 data cache are all bit-sliced into 2 parts.

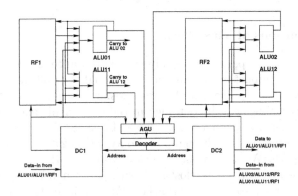

Fig. 2. Schematic 2-way Bit-Sliced Execution Core Datapath

As can be seen in Figure 2, ALU01 and ALU11 can only access RF1, and ALU02 and ALU12 can only access the RF2, and data loaded from DC1 is bypassed only to ALU01 and ALU11 and loaded only into RF1, and data loaded from DC2 is bypassed only to ALU02 and ALU12 and loaded only into RF2. The datapath for the stores is also similar. However, load and store instructions that need to load and store a single byte or a single half word, may lead to a transaction between RF1 and DC2, because the byte or the half word that is being accessed in the data cache may be present in DC2. Hence, the values from DC2 also need to be forwarded to the lower-bit ALUs and RF1, and vice versa. In our architecture, the transactions between the lower order bit slice and DC2 require an additional cycle. Since values to be stored in the cache are placed into the write buffers, the write buffers can also be bit-sliced. In our design, for simplicity, we assume that the multiplier functional unit, responsible for executing complex instructions, is not bit-sliced, because of lower frequency of complex instructions and difficulties in bit-slicing some of the complex operations. Hence, complex instructions wait till their entire operands are available, before they start execution.

In the proposed architecture, the load and store instructions cannot issue until their entire address has been computed. The advantage with the simple instructions executing on the ALUs was that the lower end bits of the operands were available for the consumer instructions even if the producer instruction had not finished execution on the upper bits. This advantage is not available for the load instructions. For simplicity, in our initial design, we assume that the AGUs wait till the entire operands are available before computing the effective address. This may delay the issue of the load and store instructions. Later in Section 4, we will see how this constraint is relaxed.

We do not bit-slice the Floating-Point (FP) subsystem[1]. However, when a value is loaded from a bit-sliced data cache, the FP instructions that are dependent on load instructions will have to wait additional cycles for their entire operands to become available. Hence, a slightly higher performance impact can be expected for the FP benchmarks. However, the simple ALU instructions dependent on the load instructions do not need to wait, because they can start executing on the lower bits when they become available and execute on the higher bits in the following cycles as and when they become available.

The execution of an integer instruction in the bit-sliced datapath takes place as follows. If an instruction is issued to the ALUs for execution, it starts execution on the lower end bits of the operands and operations on the higher end bits are performed only if required. Hence, values are read from RF2, and the result values are written into RF2 only when required. The complex instructions start execution only when all the bits of the operands are available, and these instructions write the values both in RF1 and RF2. The memory operations also compute the effective address only when the entire operands are available (delaying the issue of the load and store instructions). Once, the effective address is computed, first DC1 is accessed and then DC2 is accessed if required. When loading or storing single bytes or half-words from DC2, nothing is done in the DC1 access pipeline stage and in the next cycle DC2 is accessed.

2.2 Determining Requirement

In the previous section, we observed that the higher bit-slices of register file and the data cache are accessed only when required. To determine if a register file bit-slice needs to be read, we use an additional bit (called *next read* bit) for every register entry. When values are written into the register files, the corresponding bits are set. For instance, when a value is written into a register entry in RF2, the *next read* bit corresponding to that register entry is set. When a register entry in RF1 is read, the *next read* bit for that register entry is also read simultaneously, and depending on the value of the bit, the entry in RF2 is either read or not. RF1, storing the least significant bits, is always read. The *next read* bits are also used during concatenation of the bit-sliced values, read from all the bit-sliced register files, to form the entire operands for address computation and complex instructions. A similar procedure is employed for reading the Level 1 data caches, where a *next read* bit is employed for each word in the cache. However, the L1

---

[1] In our processor architecture, there is separate integer subsystem and a floating-point subsystem.

data caches may also be written from the lower level data cache, and for the technique to work efficiently, the *next read* bit for each word needs to be set accordingly. For this purpose, when a cache block is loaded from a lower level cache into the L1 cache, the higher end bits of all the words in the cache block are checked simultaneously and the bits for the words in the cache block are set accordingly. The *next read* bits are also used to write the correct values in the lower level cache during writeback. The number of additional bits required depends on the amount of bit-slicing done. For instance, for a 2-way bit-sliced 128-entry register file, the additional number of bits required is 128, which is only about 3% additional bits, considering 32-bit registers.

To avoid additional decoding to access the bit-vector of *next read* bits and the upper bit-slices, the decoded information from the decoder used for the lower bit-slice can be used to drive all the bit-slices and the bit-vectors. This will increase the fan-out from the decoder. However, the decoder delay is significantly lower than the delay associated with reading the register file, and the additional decoder delay can be easily absorbed in a pipelined register file access without any impact on the cycle time. The slight increase in the decoder delay can also be compensated by the reduction in the register file access time due to bit-slicing, which we confirmed using a modified version of the cacti tool [15]. A similar approach also works for the data caches.

3   Performance Results

3.1   Experimental Setup

We use the SimpleScalar simulator [3], simulating a 32-bit PISA architecture. However, we modify the simulator so that it has a separate register file, issue queue and rob, instead of a single RUU structure representing all of them. The hardware features and default parameters that we use are given in Table 1. For benchmarks, we use a collection of 7 SPEC2000 integer programs (gzip, vpr, mcf, vortex, bzip2, twolf, and gcc), and 8 SPEC2000 FP benchmarks (equake, applu, art, mgrid, mesa, ammp, apsi, and wupwise), using *ref* inputs. The statistics are collected for 500M instructions after skipping the first 500M instructions for the SPEC2000 benchmarks. We use a feature size of 0.18 μm for energy and latency measurements.

3.2   IPC Results

Figure 3 shows the IPC results for a 2-way bit-sliced configuration, compared to a non-bit-sliced configuration. Figure 3 shows that the IPC reduction with a 2-way bit-slicing is only about 5% (with a maximum of about 9% for gzip). As discussed in Section 2.3, the main reasons for a performance loss are (i) delays in the execution of load instructions, (ii) delays in the execution of complex instructions, and (iii) increase in the branch misprediction penalty. To find out the contribution of each of these parameters to the total performance loss, we measure the IPCs of a 2-way bit-sliced configuration with (i) only the loads not

Table 1. Default Parameters for the Non-bit-sliced Configuration

Parameter	Value	Parameter	Value
Fetch/Commit Width	8 instructions	*Instr. Window Size*	96 Int/ 64 FP
ROB Size	256 instructions	*Frontend Stages*	9
Phy. Register File	96 Int/ 96 FP, 1-cycle acc. lat.	*Int. Functional units*	3 ALU, 1 Mul/Div, 2 AGU
Issue Width	5 Int/ 3 FP	*FP Functional Units*	3 ALU, 1 Mul/Div
Branch Predictor	gshare 1K entries	*BTB Size*	4096 entries, 4-way assoc.
L1 - I-cache	32K, direct-map, 2 cycle latency	*L1 - D-cache*	32K, 4-way assoc., 2 cycle latency 2 read/ write ports
Memory Latency	50 cycles first chunk 2 cycles/inter-chunk	*L2 - cache*	unified 512K, 8-way assoc., 6 cycles

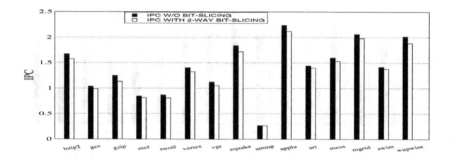

Fig. 3. IPC of a 2-way Bit-sliced Compared to a Non-bit-sliced Configuration

delayed (but the complex and branch instructions delayed), (ii) only the branch instructions not delayed, and (iii) only the complex instructions not delayed.

We found that delays in the load instructions have the most performance impact, because only the mispredicted branches impact performance, and branch misprediction rates are small for almost all the benchmarks. The percentage of complex instructions is also usually small in the benchmarks, and hence the performance impact of a delay in their execution is not significant. We also found that the increase in IPC as the load instructions are not delayed is relatively more for the integer benchmarks than the FP benchmarks. This is because, in FP benchmarks, the loads that load floating-point values still delay the dependent FP instructions, even if the address computation of the loads is not delayed.

We also used the cacti tool [15] to measure the access times of the 2-way bit-sliced register file and the data cache, and found that the access time of the data cache reduces by about 8% when going from a non-bit-sliced data cache to a 2-way bit-sliced data cache. The access time for the register file, on the other hand, reduced by a negligible amount, mainly because for a 32-bit 128-entry register file, the access time is controlled by the delay in driving the bit-lines, which does not reduce.

3.3   Energy Results

We use the cacti tool [15] to perform the energy consumption measurements for the register file and the data cache. In case of a bit-sliced cache, an access to the lower bit-slice (DC1 in Figure 2) accesses both the tag array and the data array simultaneously, and since by the time the higher order bit-slices are accessed, the cache block containing the data is already known, the access to the higher cache bit-slice (DC2 in Figure 2) occurs only to the cache block in which the data is present. Hence, the energy consumed in the higher bit-slices of the data cache is considerably less than that in the lowest bit-slice. Figure 4 shows the percentage savings in the dynamic energy consumption for the register file and the data cache.

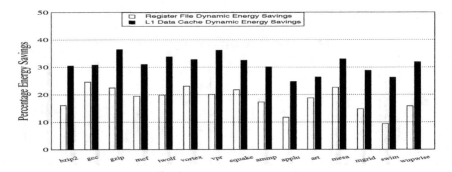

Fig. 4. Percentage Dynamic Energy Consumption Savings in Register File and Data Cache wrt Non-bit-sliced Configuration

Figure 4 shows that there is about 20% energy savings in the register file and about 30% savings in the data cache. From Figures 4 and 1(i) (we consider simple instructions because they form the majority of integer instructions), it can be seen that benchmarks that have a relatively higher percentage of instructions with operands of size greater than 16 bits have a relatively lower energy savings in the register file. This is because, for such benchmarks, the higher order register file bit-slice is also accessed frequently. For instance, consider the benchmarks applu, art, and mesa. The percentage of instructions with larger operands decreases from applu to art and from art to mesa. Correspondingly, the register file energy savings increases from applu to art and from art to mesa. In integer benchmarks as well, bzip2 has among the largest percentage of instructions with wider operands, and among the lowest percentage savings in energy consumption. Similar results are observed for the energy savings in the data cache. However, the energy savings in the FP benchmarks was observed to be less than that in the integer benchmarks, because of the wide floating-point values loaded from and stored into the data cache, which will almost always access the higher order bit-slices (as is evident from the high percentage of wide values loaded and stored for FP benchmarks in Figure 1(iv)).

4   Selective Delays

The main reasons for the reduction in IPC with a bit-sliced architecture include increase in branch misprediction penalty (due to a deeper pipeline) and delay in the issue of load instructions. In this section, we investigate techniques to recover the IPC loss due to these reasons. The basic idea is to prevent the delays from occurring, and the technique is called *selective delays*. Load instructions are delayed because they cannot be issued until the entire address is known. However, the saving grace here is that most of the effective address computations are typically performed on the lower end bits of the address and that the large base address is usually read from the register file (because once the base address is calculated it is repeatedly used with different offsets to load values from the cache). In this scenario, the AGU is designed such that if the operands are being read from the register files, all the bit-sliced register files are read simultaneously, and the effective address is computed. In this case, the load and store instructions do not get delayed. In case the large base address is being bypassed from the functional units or the data cache, then the address generation unit (AGU) waits till the entire operands are available. In this technique, the issue of instructions dependent on load instructions may have to be controlled according to whether the load instruction is reading the operand value from the register file or is receiving the operand value from the forwarding path.

To limit the increase in the branch misprediction penalty, we observe that the result of most of the branches is known after the computation in the first bit-sliced ALU. For instance, for the *branch if not equal*, if the lower end bits of the operands are different, then we know that that the branch evaluation is true irrespective of what the higher end bits are. Based on this observation, we propose that the results of the branch evaluations in the lower bit-slices of the ALUs be used (in parallel to the evaluations in the higher ALU bit-slices, which are activated if required) to detect branch misprediction and to start the recovery process as early as possible. This can avoid the increase in branch misprediction penalty for many of the mispredicted branches.

Figure 5 shows the IPCs when using the *selective delays* technique discussed in this section, and compares the IPC to that of a non-bit-sliced configuration

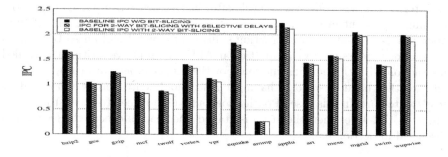

Fig. 5. Performance (IPC) Using *Selective Delays* Technique

and that of a bit-sliced configuration without *selective delays*. As can be seen in Figure 5, the IPC improves by about 3% with the *selective delays* technique, when compared to the baseline 2-way bit-sliced architecture.

5   Energy Reduction Techniques

5.1   Reducing Number of Ports

Limiting the number of register file ports has been proposed earlier by Tseng [17], in which the authors partition the register file into multiple banks where each bank contains certain registers and the number of ports in each bank is reduced. However, here we reduce the number of ports to the higher order register file bit-slices, because when bit-slicing the register file (RF) and the data cache (DC), the higher order bit-slices are not used as frequently as the lower order bit-slices. We propose that the number of read ports into the higher order RF bit-slice be reduced by half, while keeping the write ports intact for simplicity of the design. This results in a higher order FU bit-slice having only 1 read port into the higher order RF bit-slice. A single read port into the higher order RF bit-slice also works well for address generation units because they typically read only 1 32-bit operand (the base address) from the register file. If any FU requires to read two operands from the higher order RF bit-slice, the instruction that requires 2 operands from the higher order RF bit-slice, reads one operand in one cycle and then reads the other operand in the next cycle, using the same port. In this technique, the dependent instructions that may get issued in the immediately next cycle following the producer instruction, will also have to be stalled for one cycle. For this, we add another bit (called *phys-delayed*) to the *next read* bit-vector, that indicates whether the production of the higher bit-slice of a physical register will be delayed by a cycle or not. When an instruction issues, in parallel to reading the lower order RF bit-slice, it reads the *next read* bits and the *phys-delayed* bits for all the operands. If the *next read* bits of both of its operands is 1 or the *phys-delayed* bit of any of its operands is 1, it sets the *phys-delayed* bit for its destination and starts executing on the lower bit-slices of its operands (which are always available), and then it stalls for 1 cycle before continuing the execution on the higher order bits. The *phys-delayed* bit for each register needs to be *reset* 1 cycle after it has been *set*, to indicate that the ensuing dependent instructions need not wait for the higher order bits of their operands. Note that, if an instruction stalls in any bit-slice of an ALU, then no instructions are issued to that particular ALU, to avoid overwriting the stalled instruction with the new instruction. For the higher order data cache bit-slice, on the other hand, instead of having 1 read and 1 write port, we have only 1 read/write port. In this case, if both a load and a store need to access the higher order DC bit-slice, then the store is stalled and load is executed. Results are discussed in Section 5.3.

5.2   Reducing Leakage Energy Consumption

With reducing feature sizes, leakage energy consumption is becoming a significant fraction of the total energy consumption in the processor, and techniques have to be investigated to reduce the leakage energy consumption. To reduce leakage energy consumption, we investigate "shutting off" (by using *gated-Vdd* [6]) the higher bit-slices of the storage elements when they do not store bits significant to the representation of the value. For instance, for a 128-entry 32-bit 2-way bit-sliced register file, the top half of 64 registers could be shut down and during that time, the processor will have 64 32-bit registers and 64 16-bit registers. In this case, the rename logic can use *size prediction* (as discussed in [11]) to rename the instructions to appropriate registers. This technique could be applied to both the bit-sliced register file and the data caches. In this section, we only study techniques to reduce the leakage energy consumption in the register file. We explain the technique only for a 2-way bit-sliced register file with each register of 32 bits, however, it can be easily extended for further bit-sliced register files.

For a 2-way bit-sliced register file, two separate free register lists are maintained, one for the registers that have their top half "shut off" (16-bit registers) and the other for the "whole" (32-bit) registers. The size of the result produced by any result-producing instruction is predicted, and based on the prediction, an appropriate register is allocated to the instruction at rename-time. For size predictions, a *last value predictor* has been shown to work very accurately [11]. When using a *last value predictor*, the size prediction for an instruction can also be stored as an additional bit along with the instruction in the instruction cache, avoiding the use of additional tables. If no prediction can be made for an instruction, it is predicted to produce a result of size 32 bits. If a free register appropriate for the predicted size of the result is not available, then the other list is checked for free registers, and the registers are accordingly "turned on" and "turned off". "Turning on" bit-slices of registers may even take multiple cycles, and register renaming is stalled for those cycles. At the time of write-back, the size of the result is checked, and the registers are again "turned on" and "turned off" accordingly, which may again stall the pipeline for some cycles. The number of cycles the pipeline stalls depends on the number of cycles required to "turn on" the registers. As discussed earlier, if the prediction of an instruction is *more than 16 bits*, but it gets allocated a 16-bit register, then the "turn on" for the register is started at the register allocation time itself. However, by the time the instruction reaches the writeback stage, if the register is not yet completely on, the pipeline stalls for the remaining number of cycles. The stalling of the pipeline when "turning on" the registers at writeback time can have much more performance impact. The leakage energy consumption in the register file is saved by "turning off" the bits that do not store significant data. As discussed in [6], *gated-Vdd* SRAM cells (for turning off the bits) have a negligible effect on the area and the access time of the cells. Results are discussed in the next section.

5.3   Results

First, we measured the *size prediction accuracy*, and found that it hovers around the 90% mark for almost all the benchmarks. We also measured the distribution of cycles in terms of the number of 16-bit registers present in those cycles. We found that there are a considerable number of cycles that have more than 32 16-bit registers (out of a total of 96 registers). Some benchmarks (such as mesa and mgrid) even showed a considerable number of cycles with more than 64 16-bit registers.

Figure 6 shows the percentage savings in dynamic energy consumption for the register file and the data cache (with and without reduced ports) and the percentage savings in the register file leakage energy consumption. Figure 6 shows that about 5% additional energy savings are obtained in the register file when the number of read ports are reduced by half in the higher order RF bit-slice. The corresponding number for the data cache (when reducing 1 read/ 1 write port to 1 read/write port) is about 2%. The additional savings in the data cache is lower than that in the register file, because the higher order DC bit-slice as it is consumes considerably less energy than the lower order DC bit-slice (only the required cache block is accessed in the higher order bit-slice), and reduction of ports does not give significant additional energy savings. Figure 6 also shows about 20% savings in the register file leakage energy consumption. To measure the leakage energy savings, we measure the average number of register bits that are "shut off", and multiply it to the leakage energy consumption for each bit (estimated by means of Hotleakage [19]).

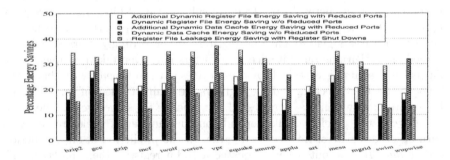

Fig. 6. Percentage Dynamic Register File and Data Cache Energy Saving (with and without Reduced Ports), and Percentage Register File Leakage Energy Saving

Figure 7 shows the IPC values with a 2-way bit-sliced configuration using reduced number of ports, and using the leakage reduction technique of "shutting off" higher RF bit-slices (with 3 different "turn on" cycle requirements of 1, 5, and 10 cycles), compared against the non-bit-sliced configuration and bit-sliced configuration with all the ports. Figure 7 shows that, when reducing the number of ports in the higher order RF and DC bit-slices (only read ports are halved for

RF), the reduction in IPC is only about 1%, compared to the baseline 2-way bit-sliced configuration with all the ports. When "shutting off" the upper bit-slices of the registers, the IPC depends on the number of cycles taken to reactivate the "shut down" registers. We assume that it takes a single cycle to shut down a register[2]. Figure 7 shows only about 3% reduction in IPC, compared to the 2-way bit-sliced configuration, when the register activation time is 1 cycle. However, for activation time of 5 cycles, the reduction is about 7% and for activation time of 10 cycles, the reduction is about 15%.

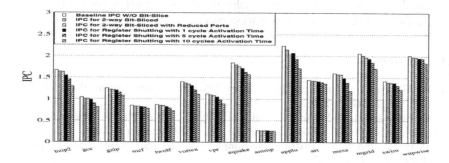

Fig. 7. Performance (IPC) of a 2-way Bit-sliced Configuration With Selective Delays; With Reduced Ports; Without Reduced Ports; and With Register Shutting

6   Conclusions

Power and performance have become two very important design criteria in the design of microprocessors. However, efforts to improve either power or performance usually leads to a degradation of the other. One important approach reduce power consumption while not hurting performance is to prevent the processor from performing unnecessary work. The techniques presented in this paper are in the same spirit, where the execution core has been bit-sliced to avoid unnecessary work. Bit-slicing uses the property that a significant amount of data in the processor is of small-size. Bit-slicing has been proposed before, but never to the extent to which we bit-slice the execution core (which includes the functional units, the register file, the upper level data cache, and the data forwarding paths). In our bit-sliced execution core, each bit-slice operates on different bits of data, and the higher order bit-slices are activated only when they are required, thus reducing energy consumption. Our studies show that, on a 32-bit machine, a 2-way bit-sliced execution core reduces the energy consumption of key hardware resources such as the register file and the data cache by about 20% and 30%, respectively, whereas the instruction throughput, with the help of performance improving techniques to prevent the instructions from getting delayed, reduces by only about 2%.

---

[2] Note that only a few registers can be shut down and activated in each cycle, thus reducing the inductive noise that can result from mass shut downs and activations.

Bit-slicing can also facilitates further reduction in the processor energy consumption. We use bit-slicing to reduce the leakage energy consumption in the register file. With reducing feature sizes, leakage energy consumption is a growing concern in the design of microprocessors. With a bit-sliced register file, we propose "shutting off" the higher bit-slices of the registers storing small-sized values, thus reducing the leakage energy consumption in them. Our studies showed that an average of about 40 registers (out of a total of 96) have their higher order bit-slices "shut off" every cycle, and that this technique reduces the leakage energy consumption in the register file by about 20%. We also propose reducing the number of ports in the higher order bit-slices to further reduce the energy consumption in the processor execution core.

References

1. A. Aggarwal and M. Franklin, "Energy Efficient Asymmetrically Ported Register File," *Proc. ICCD*, 2003.
2. D. Brooks and M. Martonosi, "Dynamically Exploiting Narrow Width Operands to Improve Processor Power and Performance," *Proc. HPCA*, 1999.
3. D. Burger and T. M. Austin, "The SimpleScalar Tool Set, Version 2.0," *Computer Arch. News*, June 1997.
4. R. Canal, A. Gonzalez and J. E. Smith, "Very Low Power Pipelines using Significance Compression," *Proc. Micro*, 2000.
5. R. Canal, A. Gonzalez and J. E. Smith, "Software-Controlled Operand-Gating," *Proc. International Symposium on Code Generation and Optimization*, 2004.
6. M. Powell, et. al., "Gated-Vdd: A circuit technique to reduce leakage in deep-submicron cache memories," *Proc. of ISLPED*, 2000.
7. M. R. Guthaus, et. al., "MiBench: A Free Commercially Representative Embedded Benchmark Suite," *Proc. IEEE International Workshop on Workload Characterization*, 2001.
8. M. K. Gowan, et. al., "Power Considerations in the Design of the Alpha 21264 Microprocessor," *Proc. DAC*, 1998.
9. G. Hinton, et al, "A 0.18-um CMOS IA-32 Processor With a 4-GHz Integer Execution Unit," *IEEE Journal of Solid-State Circuits*, Vol. 36, No. 11, Nov. 2001.
10. S. Larsen, and S. Amarasinghe, " Exploiting Superword Level Parallelism with Multimedia Instruction Sets," *Proc. PLDI*, 2000.
11. G. Loh, "Exploiting data-width locality to increase superscalar execution bandwidth," *Proc. Micro-35*, 2002.
12. S. Mahlke et. al., "Bitwidth Cognizant Architecture Synthesis of Custom Hardware Accelerators," *IEEE Transactions on Computer-Aided Design of Integrated Circuits and Systems*, 20(11), Nov. 2001.
13. G. Pokam, S. Bihan, J. Simonnet, and F. Bodin, "SWARP: A Retargetable Preprocessor for Multimedia Instructions," *Concurrency and Computation: Practice and Experience*, 16(2-3):303-318, Feb. 2004.
14. G. Pokam et. al., "Speculative Software Management of Datapath-width for Energy Optimization," *Proc. LCTES*, 2004.
15. P. Shivakumar, and N. Jouppi, "CACTI 3.0: An Integrated Cache Timing Power, and Area Model," *Technical Report, DEC Western Research Lab*, 2002.
16. M. Stepehenson et. al., "Bitwidth Analysis with Application to Silicon Compilation," *Proc. PLDI*, 2000.

17. J. Tseng, and K. Asanovic, "Banked Multiported Register Files for High-Frequency Superscalar Microprocessors," *Proc. ISCA-30*, 2003.
18. Luis Villa , Michael Zhang , and Krste Asanovic, "Dynamic zero compression for cache energy reduction," *Proceedings of the 33rd annual ACM/IEEE international symposium on Microarchitecture,* p.214-220, December 2000.
19. Y. Zhang, et. al., "Hotleakage: A Temperature-aware Model of Subthreshold and Gate Leakage for Architects," *Technical Report CS-2003-05*, University of Virginia, Department of CS, 2003.

Low-Overhead Core Swapping for Thermal Management

Eren Kursun[1], Glenn Reinman[1], Suleyman Sair[2], Anahita Shayesteh[1], and Tim Sherwood[3]

[1] Computer Science Department, University of California, Los Angeles
[2] Department of Electrical and Computer Engineering,
North Carolina State University
[3] Department of Computer Science, University of California, Santa Barbara

Abstract. Technology scaling trends and the limitations of packaging and cooling have intensified the need for thermally efficient architectures and architecture-level temperature management techniques. To combat these trends, we evaluate the thermal efficiency of the microcore architecture, a deeply decoupled processor core with larger structures factored out as helper engines. We further investigate activity migration (core swapping) as a means of controlling the thermal profile of the chip in this study. Specifically, the microcore architecture presents an ideal platform for core swapping thanks to helper engines that maintain the state of each process in a shared fabric surrounding the cores. This results in significantly reduced migration overhead, enabling seamless swapping of cores. Our results show that our thermal mechanisms outperform traditional Dynamic Thermal Management (DTM) techniques by reducing the performance hit caused by slowing/swapping of cores. Our experimental results show that the microcore architecture has 86% fewer thermally critical cycles compared to a conventional monolithic core.

1 Introduction and Motivation

Thermal characteristics of contemporary processors are creating significant challenges to microprocessor design. Various trends threaten to make things even worse: the number of on-chip transistors is quickly approaching one billion, clock frequencies are dramatically increasing, feature sizes are dropping to deep submicron levels, and supply voltage reduction is expected to slow down as it approaches noise margin barriers. As a result, power densities and on-chip temperatures are expected to increase even faster for the next generation of processors.

Thermal issues have gained significant importance in the past few years. Processor heating raises number of problems that threaten vital aspects of the microprocessor design, such as proper functionality, reliability, cost, and performance. Reliability of an electronic circuit is exponentially proportional to the junction temperature. A $10°C$ increase in temperature usually translates to $\sim2X$ difference in the lifespan of the device [16]. At higher operating temperatures the microprocessor operates at relatively lower speeds [23].

B. Falsafi and T.N. Vijaykumar (Eds.): PACS 2004, LNCS 3471, pp. 46–60, 2005.
© Springer-Verlag Berlin Heidelberg 2005

Furthermore, temperatures are not constant across the chip. $30-40°C$ thermal gradients are quite common, which causes potential timing and data errors [2]. There is a non-linear relationship between cooling capabilities and the cost of a cooling solution. The cost of cooling increases at a higher (almost exponential) rate for higher temperatures [10].

In recent years, dynamic thermal management (DTM) [4, 8, 14, 25, 15] has become an integral part of microprocessor design to adapt to increasing on-chip temperatures. The disparity between the maximum possible power dissipation and typical power dissipation has become more pronounced. This, along with the exponential increase in cooling device costs, has created a new trend where cooling systems are designed for the typical worst case power dissipation instead of the maximum possible power dissipation. Therefore, dynamic thermal management has become essential to ensure that processor temperature does not reach or exceed the maximum tolerable temperature.

Many power optimization techniques do not seem to address problems caused by processor heating, as they are targeting relatively cooler parts of the chip, such as caches. With the expected increases in power consumption and temperature, there is no doubt that more DTM techniques specific to microprocessor designs are needed.

DTM usually targets the removal of excessive heat from the processor after a certain temperature threshold is reached. Thermal management can cause performance degradation, as a result of reduced clock frequency, voltage or temporarily shutting down the entire chip. Therefore, thermal efficient architectures with less overall heating are extremely desirable, as they do not require very aggressive DTM.

In this paper we explore the thermal efficiency of the microcore architecture [18]. The microcore architecture features a small, fast pipeline augmented with helper engines [22]. All large structures are factored out of the microcore and are relocated as helper engines, taking advantage of locality in the first level structures. In this paper, we explore the use of swapping applications between multiple microcores when a given core exceeds a thermal threshold. The helper engines buffer state during core swaps and help reduce the overhead of swapping. We compare this approach to current DTM techniques.

The rest of this paper is organized as follows. In Section 2 we discuss the prior work, followed by an introduction of the architectures we investigate in Section 3. Section 4 presents the methodology. We present the experimental results in Section 5 and concluding remarks are in Section 6.

2   Related Work

The circuit design community has proposed a great deal of work on dynamic power optimization techniques, which are also used as dynamic thermal management techniques in microprocessors in various forms. Such techniques include dynamic voltage scaling (DVS) and dynamic frequency scaling (DFS). In this section we will focus on the studies that are close to our own and specifically target microprocessor power/thermal optimization.

The Pentium 4 [12] incorporates a low cost, yet reliable, thermal management system based on processor power modulation that has been commonly used in mobile systems. It utilizes the existing *stoplock*, an architectural low-power logic mechanism that halts the clock signal to the bulk of the processor [10]. Thermal management is automatically invoked whenever any of the thermal sensors indicates that the die is hotter than a predetermined critical temperature. The mechanism stays active until the die temperature drops below the critical value. The clock signal is gated at certain intervals or permanently, depending on the thermal and power management state.

Brooks and Martonosi introduced an adaptive thermal management system through speculation control in [4]. They also compared commonly used DTM techniques such as clock frequency scaling, voltage and frequency scaling, decode throttling, speculation control and instruction cache toggling [8]. An energy-management framework that combines energy efficiency and temperature management, DEETM, was presented by Huang et al. [25]. They propose several power optimization techniques such as global clock gating, DVS, sub-banking, filtered instruction cache. Although these studies provide valuable DTM techniques with significant thermal alleviation, detailed resistance-capacitance thermal models were not available at the time. As a result some of the overheating blocks were not addressed.

Him, Daash and Cai introduced a dual pipeline processor, with a secondary low-power pipeline in [15]. The power efficient single-issue, in-order pipeline only gets activated, when the primary pipeline exceed a threshold temperature. When the superscalar core overheats, it is flushed and the secondary pipeline is activated until the primary pipe cools down to a safe temperature. Register file, fetch engine and the execution units are shared among the two pipelines. However, it is important to note that this technique is mainly targeting mobile devices and applications that can tolerate low performance. There is a significant performance penalty when the architecture transitions to the secondary pipeline.

In [11] Heo, Barr and Asanovic proposed an activity migration technique for power density reduction. Activity migration reduces the temperature by moving the computation between multiple replicated blocks. This thermal reduction yields lowered leakage power values and can also be improved with a dynamic voltage scaling technique to further reduce the power and temperature.

Heo et al. [11] analyze multiple configurations with some of the microprocessor units replicated or shared. The study concludes that the best configuration has a shared Icache, Cache, rename table, and issue queue. Although, duplicated microprocessor units reduce the on-chip temperatures, they argue that this is dominated by the overhead due to activity migration.

HotSpot [14] provides an accurate thermal model and a corresponding software implementation that enables more detailed and localized thermal analysis of the microprocessor. It is based on the equivalent circuit of thermal resistance and capacitances that model the microarchitectural blocks and other aspects of the chip thermal package. Hotspot highlights the inaccuracy in estimating the temperature based on the power density only. The software models can be integrated with the other cycle accurate power estimators such as WATTCH [5] and Hotleakage [27], in order to provide a complete thermal and power analysis. In [14] Skadron et al. also provide and analyze several DTM techniques such as: temperature-tracking frequency scaling, localized toggling and computation migration. We incorporate HotSpot models for an accurate RC thermal analysis of the various architectures investigated in this study. We also make use of an idealized version of dynamic frequency scaling as a comparison point for the our core swapping approach.

3   Factored Architectures

Figure 1 illustrates a factored architecture as proposed in [18]. The main idea behind factored architectures is to move a set of larger structures out of the regular processor core, resulting in a tiny core with only the necessary components included.

While structures such as caches are fairly easy to factor, other structures require more consideration. In [18], Shayesteh et al. looked at three different types of factored structures, and their challenges:

- Hierarchical extensions: Caches and branch predictor (shown in light gray)
- Complete factorization: Value predictor and data prefetcher (shown in dark gray)
- Hybrid factorization: Register file and ROB (shown with gray stripes)

In a typical factored design, the level one data and instruction caches are moved out of the core processor pipeline and replaced with a smaller L0 cache. The L0 extends the cache hierarchy, and therefore the L1 data cache is accessed on an L0 miss.

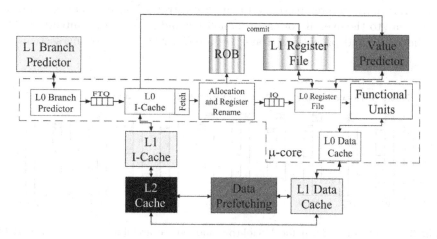

Fig. 1. The factored μ-core architecture

The architecture includes a stream buffer architecture [13] guided by a stride-filtered markov predictor as proposed in [20]. The address predictors are moved further away from the core pipeline in the microcore. There is also a hybrid value predictor [24], predicting only load instructions. To factor the value predictor, the predicted value is stored in the register allocated to the load instruction we are predicting. If the predicted value and the actual value do not match, a checker engine similar to the ARB [9] detects the misprediction and squashes the mispredicted result and its dependents.

The factored architecture makes use of a basic block target buffer (BBTB) [26], a branch address predictor that predicts an entire basic block each cycle. The microcore design has a reduced size BBTB in the core pipeline and adds a second level BBTB as done in [17]. Similarly the fetch target queue (FTQ) decouples branch prediction from the instruction cache. On a first level BBTB miss, the second level BBTB is probed and fetch stalls until a response is received from the second level. If the second level misses, we guess a fixed fetch block size and continue fetching until a misprediction is detected.

In the factored architecture, a multi-level register file is used similar to the one proposed in [3]. The basic differences are that they model an inclusive register file hierarchy where the second level register file (RF1) includes all the state contained in the first level register file (RF0). On a branch misprediction, the second level register file recovers the state of the first level register file. This is a hybrid of complete factorization and hierarchical extension, as the register file is extended with a second level structure, but the commit hardware and ROB are completely factored, with only tag allocation in the ROB impacting the core timing.

The results in [18] showed that the microcore architecture is able to reduce total processor power dissipation by 20% on average, while it attains comparable performance to a deeply pipelined monolithic design at the same clock frequency. The inherent power efficiency of the microcore, makes it an attractive design for temperature aware architectures. Figure 2 illustrates how different components contribute to the overall power for monolithic and microcore architectures. Our methodology and processor parameters are described in following sections.We

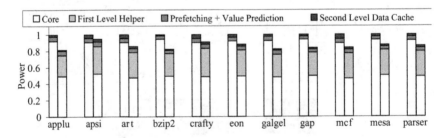

Fig. 2. Power breakdown for Monolithic and μ-core architectures. (Normalized by monolithic architecture power for each benchmark).

use the microcore framework of Shayesteh et al. to make our contribution in the analysis of the temperature efficiency and the examination of core swapping on the microcore.

3.1 Core Swapping

Swapping between multiple cores has been proposed as a dynamic thermal management technique. Heo et al. [11] look at several architectural alternatives for implementing activity migration and its overhead on processor performance. We propose a dual pipeline version of the microcore architecture, with factored components shared between the cores. Unlike [11], our core swaps are triggered by thermal sensors. When one core exceeds a thermal threshold, the application workload is swapped to the other core.

Core swapping can impact processor performance significantly. On a core swap, we flush the pipeline similar to a branch misprediction. Register file state is copied to the other core, and dirty cache blocks are written back to the level one cache (the helper engine), which is shared between the cores. We assume that copying register file state and writing back dirty blocks can be overlapped with the startup cost of the new core.

The cold start effect of caches and predictors causes an even more severe impact on the second core. These structures need to warm up and depending on their size, there is an overhead involved. In a conventional monolithic architecture, recovering from loss of data on relatively large in-core caches and predictors can degrade performance significantly. The microcore architecture, with less state in the core and more buffering between the cores, provides a very tolerant framework for core swapping. We present this feature in Section 5 by comparing the performance degradation of a monolithic core vs. a microcore in the presence of core swapping.

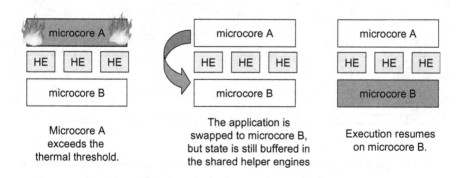

Microcore A exceeds the thermal threshold.

The application is swapped to microcore B, but state is still buffered in the shared helper engines

Execution resumes on microcore B.

Fig. 3. Core Swapping

4 Methodology

The simulator used in this study was derived from the SimpleScalar/Alpha 3.0 tool set [6], a suite of functional and timing simulation tools for the Alpha

AXP ISA. The timing simulator executes only user-level instructions. Simulation is execution-driven, including execution down any speculative path until the detection of a fault, TLB miss, or branch misprediction. Our processor operates at a 5.6 GHz clock frequency.

We used the SPEC2000 benchmark set for our experiments. Although the results are gathered for all the benchmarks, we only show results for a randomly selected subset of 6 integer and 6 floating point programs in the suite to conserve space in this paper. Details for all benchmarks will be available as a technical report (citation removed for blind review process). The programs were compiled on a DEC Alpha AXP-21164 processor using the DEC C and C++ compilers under OSF/1 V4.0 operating system using full compiler optimization (-O4 -ifo). We simulate 100 Million instructions after fast-forwarding application-specif number of instructions as proposed by Sherwood et. al in [19]. All benchmarks were simulated using the *ref* inputs.

4.1   Architectural Model

We have made significant modifications to SimpleScalar to model the various speculative techniques and different configurations in this study. We have modified SimpleScalar to include a cycle accurate, execution driven model of microcore and monolithic architecture models.

Table 1. Simulation parameters for the monolithic and microcore architectures

| | Monolithic | Microcore | |
	Core	L0	Helper Engines
Instruction Window and Physical RF	256 entry ROB 256 entry RF1	128 entry RF0	256 entry ROB 256 entry RF1
BBTB	2048-entry 4-way set associative	256-entry 4-way set associative	2048-entry 4-way set associative
L1 Data Cache	64KB 4-way set associative, dual port with a 32 byte block size, 4 cycle latency	8KB 4-way set associative, dual port with a 32 byte block size, 3 cycle latency	16KB 64-way set associative, single port with a 32 byte block size, 6 cycle latency
L1 Instruction Cache	64KB 2-way set associative, single port with a 32 byte block size, 4 cycle latency	8KB 2-way set associative, single port with a 32 byte block size, 2 cycle latency	64KB 2-way set associative, single port with a 32 byte block size, 5 cycle latency
Value Predictor (1 prediction per cycle)	2K-entry stride 8K-entry markov	none	2K-entry stride 8K-entry L2 markov
Address Predictor (1 prediction per cycle)	2K-entry stride 4K-entry markov	none	2K-entry stride 4K-entry markov
Stream Buffer	32-entry FA buffer	none	32-entry FA buffer
Branch Misprediction	26 cycles	20 cycles	
Core Width	8-way issue, 4-way decode, 4-way commit		
Memory and L2 Cache	150 cycle memory latency, 512KB 4-way set associative unified (instruction and data) cache with a 64 byte block size and 12 cycle latency		
Functional Units	8 integer ALUs, 2 integer MULT/DIV, 2 FP ALU, 2 FP MULT/DIV, 2 load/store		

Table 1 presents the simulation parameters for the monolithic and microcore architectures we explore in this paper. Cache and register file access latencies are extracted from Cacti [21] for a *70nm* Technology at 5.6 GHz frequency.

Note that the difference in branch misprediction penalty is the extra latency attributed to the larger branch predictor, register file and instruction cache in the monolithic core.

4.2   Power and Thermal Simulator

A complete analysis of the static and dynamic power consumption and resulting temperature characteristics of different architectures is crucial to our study. Our power/thermal simulator performs cycle-accurate analysis of investigated architectures based on the following recently developed power and thermal models. We used process parameters for a *70nm* process at 5.6GHz with 1V supply voltage, in order to have a better understanding of next generation submicron, low supply voltage, aggressively clocked microprocessors.

We have incorporated Wattch [5] models for dynamic power analysis of the microprocessor blocks. The experimental results we present are extracted with the most aggressive conditional clocking strategy, where the dynamic power scales linearly with access to the ports.

For submicron technologies, such as 70nm, leakage power constitutes a significant portion of the overall power. ITRS [1] predicts that leakage power is likely to increase exponentially and make up 50% of the total power dissipation for the next deep submicron processes. Hence, an accurate and reliable leakage power analysis is a necessity. We adapted leakage models from Hotleakage [27] in our power/thermal simulator. Hotleakage models are extended and improved versions of the well-known Butts and Sohi leakage equations [7]. The public version of Hotleakage only provides a software implementation of the leakage models for the data cache. We have extended and modified the tool significantly to accommodate other caches and cache-like structures in the microprocessor. We also used leakage parameters from Hotleakage's predetermined values specific to the 70nm process technology.

A detailed and accurate thermal analysis of the different architectures we explore in this study is crucial. It has been shown by [14] that thermal metrics based on power consumption or power density of individual blocks do not provide accurate thermal estimation. We used Hotspot's thermal resistance/capacitance models and RC solvers for our analysis.

Dynamic and leakage power consumption for each microprocessor unit are collected over a predetermined thermal sampling interval, as the temperatures change over periods greater than every cycle. We experimented with various sampling interval lengths, in order to explore the trade off between error rate and computational overhead. Hotspot [14] proposes a 10K instruction sampling interval for 180nm and 3.3GHz, our results showed similar error rates for 10K sampling interval for 70nm and 5.6 GHz as well.

Our power/thermal simulator also incorporates the thermal runaway phenomena enabled by Hotleakage and Hotspot models. Thermal runaway is caused by the exponential dependency of leakage power on temperature: increased temperature increases leakage power, increased leakage power causes even further increase in temperature. The positive feedback loop between leakage power and temperature is quite significant and can cause device failure.

Heo, Barr and Asanovic [11], argue that most heat is dissipated vertically on the microprocessor chip, as the wafer thickness is much smaller than the chip area. Therefore, they assume infinite lateral resistances, although it leads to the worst case temperature gradients. We follow their example, and tune HotSpot to only consider the vertical component of temperature. Lateral modeling, while possible with HotSpot, is unrealistic without a more accurate floorplan of the various architectures we consider.

Hotspot also requires a floorplan and the areas of the individual blocks of the microprocessor. We used area values based on our analysis with Cacti [21], along with a floorplan generated according to the minimum wirelength constraints. (Area values for the blocks are not presented in this version because of the page limitations.)

4.3   Dynamic Thermal Management Techniques

We assume that the critical thermal threshold is 82°C and the safety thermal threshold is 79°C for the 70nm technology process we are investigating according to the ITRS [1] projections and results from [14].

We have incorporated an idealized version of dynamic frequency scaling for the experimental analysis. Our DFS has two different frequency settings: 5.6GHz for the normal operation and 4GHz for thermal relief, which gets activated as soon as on-chip temperatures reach the 82°C critical thermal threshold. Usually there is a large latency (on the order of usecs) incurred every time the frequency is adjusted, which results in significant performance penalties in dynamic frequency scaling schemes. Skadron et al. [14] report 10usec for the non-idealized version of DFS. In our dynamic frequency scaling implementation there is no overhead, delay or penalty involved with changing the frequency of the processor.

Global clock gating is commonly used in many of todays microprocessors, such as the Pentium 4 as discussed in Section 2. We implemented a similar global clock gating mechanism for thermal analysis. The global clock signal is shut down, whenever on-chip temperatures exceed the critical thermal threshold of 82°C. The processor resumes normal operation after the chip temperatures cool down below the safety threshold of 79°C.

Our thermally-triggered core swapping mechanism gets activated as soon as a core reaches 82°C. The runs with this architecture assume an extra core (identical to the main core) that can be used to offload an application when one core overheats. The computation is migrated to the cooler core until the active core heats above the critical thermal threshold and another swap is required. Thermally-triggered core swapping minimizes the swapping overhead relative to approaches that swap at fixed intervals regardless of core temperature.

5   Experimental Results

In this section we evaluate the performance of the microcore architecture alone and in the presence of different DTMs. In particular, we examine the ability of the microcore to buffer state when core swapping, and compare this to a conventional monolithic architecture.

5.1   Thermal Characteristics of Microcore vs. Monolithic

Figure 4 compares the performance and thermal behavior of a conventional monolithic core and the microcore architecture on some of the SPEC 2000 benchmarks. The upper half of the figure shows performance in BIPS for different benchmarks, and the lower half illustrates the heating behavior of the investigated architectures. This latter component shows the percentage of cycles for which at least one block exceeds the indicated temperatures: 75°C, 79°C, 82°C and 85°C. Darker colors in the lower graphs indicate higher temperatures. The rest of the figures in this section are similarly constructed.

For example, `galgel` sees comparable performance with either the microcore or monolithic architecture, but the monolithic core sees a temperature greater than 85°C almost 97% of the time. The microcore only exceeds 85°C around 18% of the time, and stays below 82°C around 42% of the time.

Note that for many benchmarks, and particularly in monolithic architectures, temperature frequently exceeds the thermal threshold, 82°C. These results should be considered as an upper bound for performance that are not be achievable without some form of thermal management. On-chip temperatures for the microcore architecture are significantly lower than the monolithic core, but it still retains good performance comparable to that of the monolithic core. This can be attributed to the significantly smaller structures in the microcore that are much more power efficient.

Our detailed thermal analysis considers all of the possible overheating blocks. Although some of the hotspots were common among different benchmark, such as the register file, load-store queue, etc, others varied across the different bench-

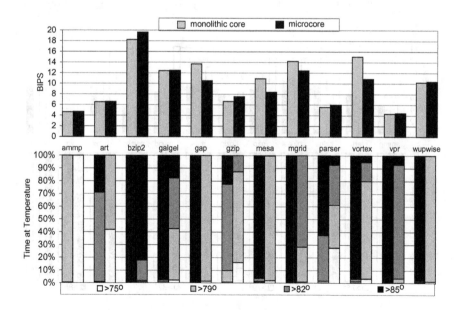

Fig. 4. Performance and thermal behavior of a microcore vs a monolithic core

marks and configurations. Even though the location of hotspots can provide a level of insight, the thermal behavior of the architecture can also be captured by the number of cycles that any of the blocks exceed a given thermal threshold.

The smaller structures of the microcore consume less power on each access compared to larger blocks in the monolithic architecture. Moreover, the larger helper engines are not accessed as frequently. Their inherent latency tolerance provides opportunities for power optimization. The microcore architecture shows performance comparable to the monolithic core, but with a 20% reduction in power on average.

It is important to note that the ITRS projects a reduction in maximum permitted junction temperatures for the future generations of process technologies. The maximum tolerated junction temperatures are around 85°C for 130nm and even lower for smaller process technologies.

The inherent thermal efficiency of the microcore also enhances the effective temperature reduction when used with DTM techniques. Next, we evaluate the performance and thermal behavior of DTM techniques, including core swapping, on the monolithic core and microcore.

5.2   Dynamic Thermal Management on Monolithic Architecture

Figure 5 shows core swapping results compared to no DTM, global clock gating, and an idealized version of dynamic frequency scaling on the monolithic architecture. The upper section of the graph displays performance in BIPS, the

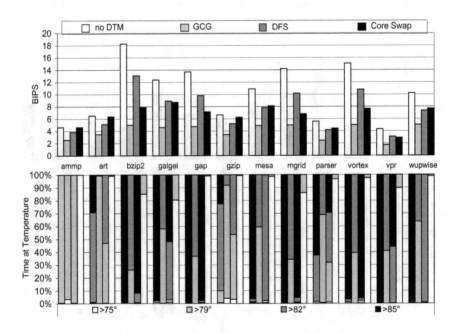

Fig. 5. DTMs on the monolithic architecture

lower part is dedicated to the thermal behavior of the same benchmark and DTMs, similar to the previous figure. Darker shades in the lower part of the figure indicate higher temperatures as well.

Core swapping results are shown in black bars (at the top part of the Figure), idealized dynamic frequency scaling in dark gray and global clock gating are in light gray. White bars demonstrate results without thermal management of any kind, no-DTM. As mentioned earlier in Section 1, performance degradation is commonly experienced with dynamic thermal management techniques. The degradation usually comes from various sources such as frequency decrease, voltage reduction, clock gating. Performance degradation might be quite significant depending on the DTM technique.

As a result no-DTM has the best performance results in BIPS among all cases. However, it is almost impossible to achieve comparable performance in reality since it would require sustained operation at a temperature beyond the critical thermal threshold, and a processor operating under such conditions would likely have timing, data and reliability complications. Although global clock gating seems to be more effective in reducing the temperature in most benchmarks than DFS, it has a very significant performance penalty as a result of disabling the global clock signal frequently.

Core swapping is extremely effective at thermal management, reducing the temperature below 79°C at least 80% of the time for all benchmarks and well above 95% of the time for many benchmarks. On the monolithic core, some applications are able to tolerate the performance impact of core swapping, but there is a pronounced degradation for many benchmarks, like bzip2 and mgrid.

For the monolithic case, temperatures were still above the threshold for many applications with DFS, such as bzip2, gap and mgrid. This may indicate that our DFS strategy requires an even lower frequency to provide thermal relief to these applications, but at an even greater cost to performance. Despite a 70% drop in performance mgrid is still above 85°C around 95% of the time with DFS. gap operates in lower frequency mode almost 99% of the time in order to reduce the temperature, yet it is still above the 85°C temperature threshold 97% of the time.

5.3   Dynamic Thermal Management on Microcore

Figure 6 shows the behavior of the microcore with DTM techniques. We observe significantly improved thermal behavior compared to the monolithic architecture (Figure 5), and see less performance degradation from core swapping.

It is important to note that state buffering provided by the shared helper engines minimizes the core swapping overhead in the microcore architecture. Core swapping is always able to outperform the other DTMs on a microcore architecture, in most cases coming close to the performance of the architecture without any DTM. It has an equally dramatic impact on temperature in the microcore architecture. Temperatures are lower than 82°C with core swapping, for all of the benchmarks. Even galgel, which spends over half its execution time over 82°C is able to reduce its temperature below 79°C around 93% of the time using core swapping, with only an 8% degradation in BIPS.

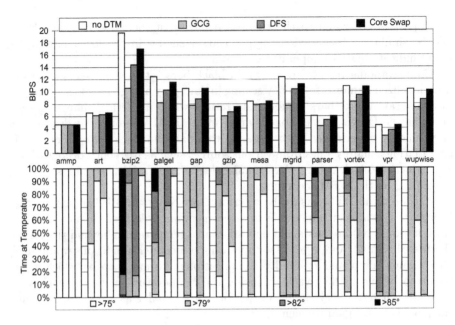

Fig. 6. DTMs on the microcore architecture

Note that we have used an idealized DFS implementation (see Section 4). This behavior can cause a significant performance degradation if frequency switching is used often. Notice that the idealized DFS is given competitive advantage against a core swapping approach with a realistic performance penalty. Despite this, core swapping is still able to outperform DFS.

6   Summary

In this paper, we investigated the thermal behavior of the microcore architecture, and examined the use of core swapping as a legitimate alternative to conventional DTMs.

We demonstrated that the microcore architecture enables lower on-chip temperatures compared with a conventional monolithic architecture. Factoring large, power-hungry units out of the core limits the number of accesses to such blocks and prevents them from heating as much. Our experiments show that the microcore reduces number of cycles over the critical thermal threshold by 86% on average, even without any thermal management use.

Furthermore, we have proposed a thermally-triggered core swapping mechanism as a dynamic thermal management technique. Microcores enable efficient core swapping by buffering processor state in shared helper engines that reduce startup costs when switching to a new core. Our experimental results indicate that a microcore is able to attain comparable IPC to a monolithic core, but with 94% fewer cycles above the critical thermal threshold.

The core swapping mechanism shows promising thermal reduction ability. It does not suffer any cycles in thermal violation for any of the benchmarks we examined. It also has favorable performance (as measured in BIPS) when compared to other DTM techniques such as GCG and the idealized DFS.

Future microprocessor generations have great thermal challenges awaiting them. Thermally efficient architectures and dynamic thermal management techniques are both critical to overcoming these challenges. Architectures like the microcore can help to achieve this without sacrificing performance.

References

1. In *International Technology Roadmap for Semiconductors*, 2003.
2. A. Ajami, K. Banerjee, M. Pedram, and L. van Ginneken. Analysis of non-uniform temperature-dependent interconnect performance in high performance ics. In *41st Design Automation Conference*, pages 567–572, June 2001.
3. R. Balasubramonian, S. Dwarkadas, and D. Albonesi. Reducing the complexity of the register file in dynamic superscalar processors. In *Proceedings of the 34th Annual International Symposium on Microarchitecture*, December 2001.
4. D. Brooks and M. Martonosi. Adaptive thermal management for high-performance microprocessors. In *Workshop on Complexity Effective Design*, June 2000.
5. D. Brooks, V. Tiwari, and M. Martonosi. Wattch: A framework for architectural-level power analysis and optimization. In *27th Annual International Symposium on Computer Architecture*, pages 83–94, June 2000.
6. D. C. Burger and T. M. Austin. The simplescalar tool set, version 2.0. Technical Report CS-TR-97-1342, U. of Wisconsin, Madison, June 1997.
7. J.A. Butts and G.S. Sohi. A static power model for architects. In *27th Annual International Symposium on Computer Architecture*, pages 191–201, June 2000.
8. D.Brooks and M.Martonosi. Dynamic thermal management for high-performance microprocessors. In *International Symposium on High-Performance Computer Architecture (HPCA-7)*, pages 171–182, January 2001.
9. M. Franklin and G. S. Sohi. Arb: A hardware mechanism for dynamic reordering of memory references. *IEEE Transactions on Computers*, 46(5), May 1996.
10. S. Gunther, F. Binns, D. Carmean, and J. Hall. Managing the impact of increasing microprocessor power consumption. In *Intel Technology Journal Q1*, 2001.
11. S. Heo, K. Barr, and K. Asanovic. Reducing power density through activity migration. In *International Symposium on Low Power Electronics and Design*, August 2003.
12. G. Hinton, D. Sager, M. Upton, D. Boggs, D. Carmean, A. Kyker, and P. Roussel. The microarchitecture of the pentium 4 processor. *Intel Technology Journal Q1*, 2001.
13. N. Jouppi. Improving direct-mapped cache performance by the addition of a small fully associative cache and prefetch buffers. In *Proceedings of the 17th Annual International Symposium on Computer Architecture*, May 1990.
14. K.Skadron, M.Stan, W. Huang, S.Velusamy, K. Sankaranarayanan, and D. Tarjan. Temperature-aware microarchitecture. In *30th Annual International Symposium on Computer Architecture*, pages 2–13, June 2003.
15. C-H. Lim, W. Daasch, and G.Cai. A thermal-aware superscalar microprocessor. In *International Symposium on Quality Electronic Design*, pages 517–522, March 2002.

16. L.T.Yeh and R.Chu. Thermal management of microelectronic equipment. In *American Society of Mechanical Engineers - ISBN:0791801683*, 2001.

17. G. Reinman, T. Austin, and B. Calder. A scalable front-end architecture for fast instruction delivery. In *26th Annual International Symposium on Computer Architecture*, May 1999.

18. A. Shayesteh, E. Kursun, S. Sair, T. Sherwood, and G. Reinman. An evaluation of deeply decoupled cores. In *University of California Los Angeles Tech Report CS-2004-09*, 2004.

19. T. Sherwood, E. Perelman, G. Hamerly, and B. Calder. Automatically characterizing large scale program behavior. In *Proceedings of the 10th International Conference on Architectural Support for Programming Languages and Operating Systems*, October 2002.

20. T. Sherwood, S. Sair, and B. Calder. Predictor-directed stream buffers. In *33rd International Symposium on Microarchitecture*, December 2000.

21. P. Shivakumar and Norman P. Jouppi. Cacti 3.0: An integrated cache timing, power, and area model. In *Technical Report*, 2001.

22. J. E. Smith. Instruction-level distributed processing. *IEEE Computer*, 34(4):59–65, April 2001.

23. R. Viswanath, V. Wakharkar, A. Wathe, and V.Lebonheur. Thermal performance challenges from silicon to systems. In *Intel Technology Journal Q3*, 2000.

24. K. Wang and M. Franklin. Highly accurate data value prediction using hybrid predictors. In *30th Annual International Symposium on Microarchitecture*, pages 281–290, December 1997.

25. W.Huang, J.Renau, S-M.Yoo, and J. Torrellas. A framework for dynamic energy effiency and temperature management. In *33rd International Symposium on Microarchitecture*, pages 202–213, December 2000.

26. T. Yeh and Y. Patt. A comprehensive instruction fetch mechanism for a processor supporting speculative execution. In *Proceedings of the 25th Annual International Symposium on Microarchitecture*, pages 129–139, December 1992.

27. Y. Zhang, D. Parikh, K. Sankaranarayanan, K. Skadron, and M. Stan. Hotleakage: A temperature-aware model of subthreshold and gate leakage for architects. In *University of Virginia Dept of Computer Science Tech Report CS-2003-05*, March 2003.

Software–Hardware Cooperative Power Management for Main Memory*

H. Huang[1], K.G. Shin[1], C. Lefurgy[2], K. Rajamani[2], T. Keller[2],
E. Hensbergen[2], and F. Rawson[2]

[1] The University of Michigan, Ann Arbor, MI 48105, USA
{haih, kgshin}@eecs.umich.edu
[2] IBM Austin Research Laboratory, Austin, TX 78758, USA
{lefurgy, karthick, tkeller, bergevan, frawson}@us.ibm.com

Abstract. Energy is becoming a critical resource to not only small battery-powered devices but also large server systems, where high energy consumption translates to excessive heat dissipation, which, in turn, increases cooling costs and causes servers to become more prone to failure. Main memory is one of the most energy-consuming components in many systems. In this paper, we propose and evaluate a novel power management technique, in which the system software provides the memory controller with a small amount of information about the current state of the system, which is used by the memory controller to significantly reduce power. Our technique enables the memory controller to more intelligently react to the changing state in the system, and therefore, be able to make more accurate and more aggressive power management decisions. The proposed technique is evaluated against previously-implemented power management techniques running synthetic, SPECjbb2000 [17] and various SPECcpu2000 [18] benchmarks. Using SPEC benchmarks, we are able to show that the cooperative technique consumes 14.2–17.3% less energy than the previously-proposed hardware-only technique, 16.0–25.8% less than the software-only technique.

1 Introduction

This paper focuses on reducing power dissipated by the main memory system (consists of DRAM). This is motivated by a continuous increase in the power budget allocated to the memory. For example, as much as 40% of the system energy is consumed by the memory system in a mid-range IBM eServer machine [11]. Power dissipated by the DRAM is largely dependent on its capacity and bus frequency. Therefore, as applications become increasingly data-centric, for the performance of the system to continue to scale, we would need more power to sustain a larger-capacity and higher-performance memory system, which can easily dominate the total system energy budget.

The main contributions of this paper are summarized as follows.

- Design of a novel power management technique that enables the system software to cooperate with the memory controller hardware by providing it with critical system-state information which was previously unavailable at the hardware level. This

* The work reported in this paper was supported in part by the US Air Force Office of Scientific Research under Grant AFOSR F49620-01-1-0120 and also by DARPA under Contract No. NBCH3039004.

B. Falsafi and T.N. Vijaykumar (Eds.): PACS 2004, LNCS 3471, pp. 61–77, 2005.

allows the memory controller to more intelligently react to the changing state in the system, and therefore, significantly improve the energy-performance efficiency of main memory.

- Use of a full system simulator (Mambo [4]) and a systematic evaluation methodology to accurately simulate the behavior of the proposed power management unit in the memory controller and its performance and energy effects on the system. Using a modified 2.6.5 Linux kernel, it enables us to precisely identify problems and benefits associated with the proposed cooperative management technique running various types of workload.
- Evaluation of registered DRAM (server-grade), which has been mostly underexplored in the past, but it is now becoming increasingly important as it is almost always used in today's server systems. Using registered DRAM, we demonstrate that our cooperative technique can save 14.2–17.3% more energy than previously-proposed hardware-only techniques, 16.0–25.8% more than software-only techniques, and 71.6–75.8% more than no power management.

The rest of the paper is organized as follows. Section 2 provides background information on the current state of DRAM technology and various DRAM architectures. Section 3 describes the detail in the proposed cooperative technique which consists of (i) Power Aware Virtual Memory (PAVM) implemented in the OS, (ii) a thin power management layer in the memory controller hardware, and (iii) a software-hardware interface. Experimental setup and detailed evaluation are given in Section 4, demonstrating a significant benefit in using this new approach in terms of energy and performance. Section 5 discusses related work, and Section 6 highlights some future research directions and finally concludes the paper.

2   Memory System Model

In this section, we discuss performance and energy implications when power is managed for the main memory. Since 1980, the performance gap between the memory and the processor has been widening continuously — DRAM speed has been only improving at an annual rate of 7% while processor speed has been improving at an annual rate of 40% [19]. Furthermore, frequent interaction between memory and other system I/O components makes it a crucial component in the overall performance of the system. Unfortunately, power reduction is only possible when memory is operating at lower performance states, and therefore, it is important to ensure that either this performance degradation can be hidden or that the energy saved in the memory justifies the performance degradation that it causes. In this paper, we mainly concentrate on DDR as it is becoming the most-widely used memory type. Nevertheless, our technique is architecture-independent and can be easily applied to other memory types.

2.1   Double-Data Rate DRAM Model

DDR memory is usually packaged as modules, or DIMMs, each of which usually contains either 1, 2 or 4 *ranks*, which are commonly composed of 4, 8 or 16 number of physical devices Each time a DIMM is accessed, 64 bits of data is read or written.

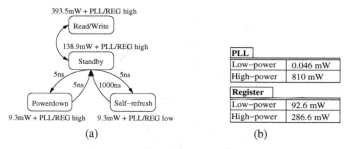

Fig. 1. (a) Power dissipated in each power state and the delays to transition between these states for a single 512-Mbit DDR device. (b) Power dissipation of a TI CDCVF857 PLL device (one per DIMM) and a TI SN74SSTV32867 register.

Since each device, depending on design, can supply either 4, 8, or 16 bits at a time, multiple devices are needed to act simultaneously to satisfy a 64-bit DIMM access, and these devices constitute a rank. A rank is then divided into multiple banks (logical devices, usually 4 or 8), each of which may be accessed individually, but cannot be power managed separately. The smallest physical unit for which we can independently manage power is a single rank.

DDR architecture has many power states defined and even more possible transitions between them [15,10]. For simplicity of presentation, we only consider four of these power states — Read/Write, Standby, Powerdown, and Self Refresh — listed in a decreasing order of power dissipation. The power dissipation in each state and the transitional delays between them are shown in Figure 1(a). Note that the power numbers shown here are for a single device. Therefore, to calculate the total power dissipated by a rank, we need to multiply this power by the number of devices used per rank. For a 512MB registered DIMM consisting of 8 devices in a rank, the expected power draw values are 4.2 W, 2.2 W, 1.2 W, and 0.167 W, respectively, for the four power states considered here. The details of these power states are as follows:

- **Read/Write:** Dissipates the most power, and it is only briefly entered when a read/write operation is in progress.
- **Standby:** When a rank is neither reading nor writing, Standby is the highest power state, or the most-ready state, in which read and write operations can be initiated immediately at the next clock edge.
- **Powerdown:** When this state is entered, the input clock signal is gated except for the refresh signal. I/O buffers, sense amplifiers and row/column decoders are all deactivated in this state.
- **Self refresh:** In addition to all the components on a DIMM that are deactivated in Powerdown, the phase-lock loop (PLL) device and registers are also put to the low-power state to maximize energy savings as the PLL and the registers (Figure 1(b)) can consume a significant portion of the total energy on each DIMM. However, when exiting Self Refresh, a 1 μsec delay is needed to re-synchronize both the PLL and the registers.[1]

---

[1] Registered memory is almost always used in server systems to better meet timing needs and provide higher data integrity, and the PLL and registers are critical components to take into account when evaluating registered memory in terms of performance and energy.

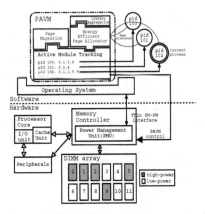

Fig. 2. Architectural overview of cooperative power management system

Due to having a large power differential between Standby and Powerdown / Self Refresh, we want to minimize the time a rank stays in Standby and maximize the time it spends in either Powerdown or Self Refresh. However, at the same time, we also want to minimize performance degradation caused by accessing ranks that were previously put to one of the low-power states. Therefore, determining which ranks to power down, when to power down, and into which low-power state to transition are critically important to both energy and performance. For the time-being, we refer to Standby as the high-power state, and both Powerdown and Self Refresh as low-power states. We make the distinction between these two low-power states in Section 4 and illustrate how to best utilize each to maximize energy savings while minimizing performance impact.

3   Design

This section details the design of the cooperative power management technique. It begins with a brief design overview in Section 3.1. Hardware and software-side control mechanisms are described in Section 3.2 and Section 3.3, respectively.

3.1   Overview

Proposals to manage power in the memory system have traditionally operated solely in the hardware domain [2,5] or in the software domain [6,9], but not in both. In our work, we discovered that a small amount of cooperation between these two domains can lead to a significant energy benefit. In the hardware-controlled power management approach, memory traffic is monitored by the memory controller which permits implementation of a very fine-grained and highly-adaptive control mechanism, which ideally can be used to glean all possible energy-saving opportunities. However, the effectiveness of this approach is usually limited by how well the hardware can predict future references from the past access behavior. Accurate prediction is very difficult to accomplish at such a low level, especially in a complex multitasking system, where the memory access patterns constantly change due to interleaved execution of many different processes.

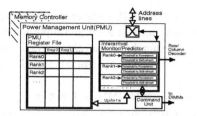

Fig. 3. Architecture of a per-rank PMU implemented in the memory controller

Any incorrect prediction will translate into both performance and energy penalties. On the other hand, in the software-controlled, or more precisely OS-controlled, approach, system and process state information (e.g., which memory regions are used by which process) can be easily tracked by the system software. This information then enables the OS to avoid performance penalty when managing the power for the memory as it can keep all ranks that may be used by the current running process in a high-power ready state while having all other ranks in low-power states. However, system software alone is not capable of achieving fine-grained power control, as the OS is not generally aware of which ranks a process is *accessing* at run-time, or how actively it is accessing a rank, or whether or not there are any memory access patterns that can be exploited. It only knows about the active ranks of the running process. However, since some of the active ranks are infrequently used, and due to its inability of exploiting such knowledge, many energy-saving opportunities will be lost in using this software-only approach.

Based on this observation and our discovery of a complementary relationship between these two types of approaches, we propose a cooperative power management approach that exploits the unique features available in each domain that can be used to aid the other. For example, fine-grained control mechanisms available in the hardware level can be used to aid the system software to re-capture some of the missed energy-saving opportunities described earlier. Conversely, the system software can export useful system and process state information down to the memory controller, so that the observed memory traffic can be better interpreted at the hardware level, thus allowing the hardware to make more accurate power management decisions. Figure 2 depicts the system architecture of this cooperative power management approach showing both the software and hardware components. In the next section, we first describe the architecture of the power management unit (PMU) in the memory controller. It is the hardware component responsible for monitoring memory traffic and controlling power in the DRAM. We then describe how to minimally modify this PMU so it can efficiently communicate with the system software to gain information about the current state of the system, and thus allowing it to more intelligently manage power. In Section 3.3, we describe what system and process state information are useful to the PMU and how does the system software convey this information to the PMU.

3.2 Context-Aware PMU

Memory-controller-based power management [2,7,8] has been previously proposed to provide fine-grained monitoring and power control, which is usually performed by a separate power management unit (PMU) implemented within the memory controller. This PMU is typically implemented as a set of simple logic devices that (i) monitor main memory accesses, (ii) predict threshold values to determine when to power down, and (iii) instruct the memory controller to perform power-down operations when certain conditions are met.

A schematic diagram of a simple PMU is shown in Figure 3. It monitors memory accesses by snooping the address lines and keeps track of the past access behavior in an internal register file, where the number of registers is dependent on how accurate we need the prediction logic to be. Based on the history, a threshold value is derived to determine how much idle time should elapse before putting a rank into a low-power state. When multiple energy-saving states are implemented, one can derive multiple thresholds, each used to transition the rank to a different low-power state.

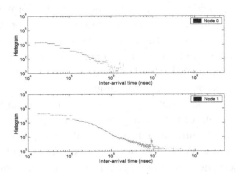

Fig. 4. Inter-arrival time observed on two different ranks (or nodes)

Separate monitor/predictor logic is often kept for each of the ranks so the PMU can individually monitor memory accesses, keep history and control power state for each. The reason for keeping a separate set of logics is because each rank may be accessed very differently from all other ranks. To give an example, Figure 4 shows a histogram (in log scale) of inter-arrival times (in log scale) between consecutive memory accesses observed on two different ranks. It is apparent from this figure that the access characteristics observed on these two ranks are very different. On rank 0, we can observe that with most inter-arrival times being very short, nearly every memory access comes within 1 msec after the previous one. On rank 1, however, there are many larger gaps (indicated by a heavier-tailed distribution) between memory accesses, suggesting that we have more energy-saving opportunities and also the fact that different thresholds should be used on these two ranks to maximize energy savings on each. However, this *per-rank* implementation in the PMU would require additional circuitry which not only adds manufacturing costs but also additional energy costs. Later, we will show how to use the process state information exported by the system software to reduce this additional cost.

Per-Process Power Management. In the previous section, we illustrated the mechanism to monitor memory traffic and manage power on a per-rank basis. Now, to take this concept a step further in enabling the controller to better interpret the monitored

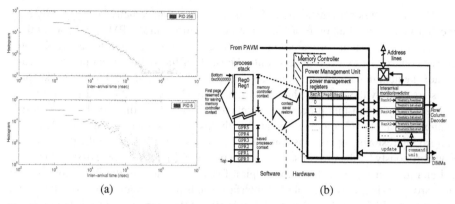

Fig. 5. (a) Inter-arrival time incurred by two different processes observed on the same memory rank. (b) Architecture of the Process-Aware PMU in the memory controller.

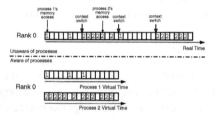

Fig. 6. An example that gives some intuition on why it is beneficial to make the memory controller context-aware

memory traffic, we further partition the observed per-rank memory traffic on a *per-process* basis. The reason why this is important is that different processes can exhibit vastly different memory access behaviors. Even for processes with similar access behaviors, how they access each individual rank can be quite different (Figure 5(a)) given that the virtual-to-physical page mapping is controlled arbitrarily by the OS. So, if the PMU has no understanding of processes, the observed per-rank memory traffic is essentially "polluted" by all processes that access this rank in rapid successions (at a 10 msec or even an 1 msec quantum) as scheduled by the task scheduler. Therefore, the PMU will likely make inefficient power management decisions based on this "average" access behavior observed from all the concurrent processes. We illustrate this by an example shown in Figure 6. In this example (top portion of the graph), Process 1 rarely accesses rank 0, whereas Process 2 accesses this rank very frequently. If the controller monitors the memory traffic on this rank without differentiating between the two processes, it will conclude that this rank is accessed "moderately", and thus, might make less-than-optimal power management decisions. However, by making the memory controller *context-aware* (bottom portion of the graph), the PMU can easily detect that Process 1(2) rarely(frequently) accesses this rank, and therefore, can select more suitable thresholds depending on which process is currently executing. The problem, however, is that unlike in the case of per-rank management, the memory controller is totally oblivious to the concept of a process, which ironically strongly impacts how the memory is being accessed and how it should be controlled.

The improvement to make the PMU context-aware can actually be very easily augmented with a small amount of hardware modifications in the PMU and some minor changes to the system software. On the software side, in addition to saving the processor context (i.e., CPU registers) onto the stack of the switched-out process at each context switch, in parallel, we would also need to save the values of the history-keeping registers used by the PMU as shown in Figure 5(b) (Ignore the `PAVM line` for now). Subsequently, when this process is switched back at a later time, both the processor context and the PMU context associated with this process are restored. The PMU context saving/restoring operations can either be done synchronously by the processor, or asynchronously by the PMU itself when the processor sends it a context-switching signal and gives it a physical memory region for saving/restoring the PMU context. On the hardware side, only a simple I/O interface needs to be implemented for saving and restoring the PMU context. Essentially, this allows the memory controller to more efficiently manage power for the main memory tailored to the memory access behavior specific to each process because the PMU can now make power management decisions solely based upon each process's past memory access behavior.

3.3 Interactions with PAVM

Power Aware Virtual Memory (PAVM) was first proposed and implemented by Huang *et al.* in [9]. It leverages OS-level information and can make very accurate power management decisions, thus only negligibly affecting performance when performing power management. We discovered that the information collected by PAVM in the operating system can be used by the PMU to make more accurate power management decisions and to determine which monitor/predictor circuits in the PMU are unnecessary so it can turn them off to further reduce power. The availability of a full-system simulator enables us to find several problems with the original PAVM implementation. We found that a small but a non-negligible number of memory accesses did not go to the active ranks of the current running process. These were later found to be memory accesses incurred by the kernel (i.e., through system call, interrupt, exception) while in user process's context. This was resolved by tagging all pages that are used only by the kernel and aggregating them onto the first rank in the system and always keeping this rank in the most-ready state to reduce performance impact. In our experiment, a single 64MB memory rank seems to have more than enough capacity for such purpose.

As indicated in Section 3.2, even though only a small amount of modifications is needed to implement the aforementioned energy-conserving mechanisms in the hardware, but the additional hardware does not come for free — a small but a non-negligible additional power is dissipated. To amortize this cost, PAVM can inform the PMU which ranks are used by the running process so that the PMU can completely gate off all the monitor/predictor circuits and history-keeping registers for those inactive ranks without affecting the effectiveness of the power management mechanism. This information is passed down from the `PAVM control` line shown in Figure 5(b).

Cooperations with PAVM also have certain performance benefits. So far, we have only discussed policies and mechanisms to power down ranks but not to power them up. As premature power-ups waste energy, we currently do not consider any power-up heuristics in the hardware. Instead, we rely on a simple but accurate power-up mechanism implemented in PAVM. Since many memory accesses occur immediately after a context switch due to cold cache misses, if PAVM can instruct the memory controller to power up the active ranks of the to-be-run process as early as possible, some re-synchronization penalties can be avoided.

4   Evaluation

We now evaluate the effectiveness of the proposed cooperative HW–SW power management technique and compare it against some previously-proposed techniques. Section 4.1 describes the simulation environment and the methodology that we have used to collect and analyze results. Section 4.2 and Section 4.3 provide detailed simulation results using synthetic and SPEC benchmarks (SPECjbb2000 and SPECcpu2000), respectively.

4.1   Simulation Setup

To the best of our knowledge, the proposed PMU architecture is not available in any commercial systems to date. Therefore, the best one can do is to use a machine simulator; we choose to use Mambo [16] in this project. Mambo is a full-system simulator for PowerPC® machine architectures and is in active use by multiple research and development efforts at IBM. It emulates both 32-bit and 64-bit PowerPC® processors and also supports various system architectures and components, including a multi-tiered cache hierarchy, SLBs, TLBs, disks, Ethernet controllers, UART devices, etc. We used a modified 2.6.5-rc3 Linux kernel, running on top of a Mambo simulated machine (parameterized as shown in Table 1) to run all our workloads.

To evaluate various power-management techniques, we first use Mambo to record all the main memory traffic (i.e., filtered by the L1 and L2 caches) into a trace file, and then feed it into a trace-driven main memory simulator to simulate various power-management decisions that could have been made by the memory controller at runtime. This memory simulator is written using CSIM [13] library, and it can simulate detailed activities in memory devices, controllers, synchronous memory interfaces (SMIs) and on various buses. Instantaneous power is calculated using the method described in [14]. We keep track of state information for each bank on a per-cycle basis, which gives us power and performance information.

4.2   Synthetic Benchmark

We first use a synthetic benchmark consisting of two streaming processes. The first process's memory accesses all miss in the cache and go to the main memory, and the second process's all hit in the cache. This synthetic benchmark is not meant to be realistic, but through this simple example, we can illustrate the potential benefit in making the memory controller context-aware. Furthermore, using this simple scenario, we can also see more clearly what are the energy and performance implications in using various power management techniques. In the following section, we evaluate and compare these power management techniques with more realistic workloads — SPECjbb2000 and SPECcpu2000.

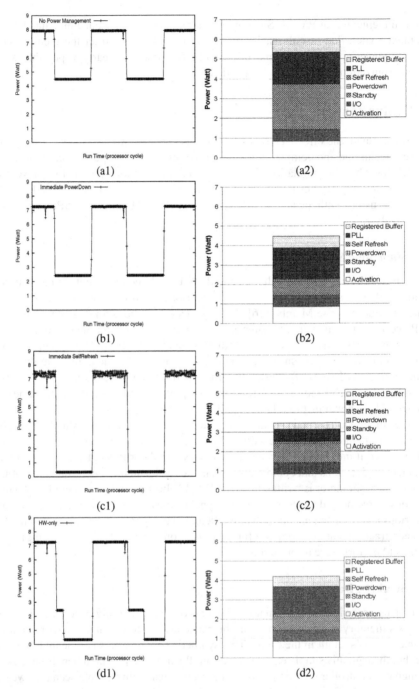

Fig. 7. The first column shows the instantaneous power for a zoomed-in period of the synthetic workload under (a) no power management (b) Immediate Power Down (c) Immediate Self Refresh (d) HW-only and (e) HW–SW techniques. The second column shows the breakdown of the average power dissipated.

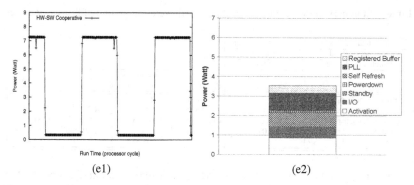

(e1)   (e2)

Fig. 7. (*Continued*)

Table 1. System parameters used in Mambo. All cache lines are 128 Bytes long.

Component	Parameter
Processor	64-bit 1.6GHz PowerPC®
DCache	64KB 2-way Set-Associative
ICache	32KB 4-way Set-Associative
L2-Cache	1.5MB 4-way Set-Associative
DTLB	512 entries 2-way Set-Associative
ITLB	512 entries 2-way Set-Associative
DERAT	128 entries 4-deep
IERAT	128 entries 4-deep
SLB	16 entries
Memory	DDR-400 768MB (64Mbx8)
Linux Kernel	2.6.5-rc3 w/ PAVM patch

Power Management Techniques. The machine configuration used for this benchmark is the same as that shown in Table 1, except that the memory capacity is reduced to a single 64MB rank. The two streaming processes are scheduled in an interleaved-manner by the Linux task scheduler. Without any power management, the instantaneous power dissipated by the memory is shown in Figure 7(a1), where one can clearly see when each process is scheduled. In Figure 7(a2), we break the average power dissipated for this benchmark down to various components. Power used by activation, read, write operations and data queues are due to DRAM devices doing useful work and cannot be reduced by using power management. Here, we look for ways and opportunities to reduce the idle power that is wasted when no work is done. Most of this idle power is dissipated in the Precharge Standby mode, Active Standby mode, and by the PLL and the registers.

First, we consider the simplest static hardware techniques, which try to put the rank to either Power Down or Self Refresh mode immediately at the end of each memory request. We call them Immediate Power Down and Immediate Self Refresh, respectively, and the results are shown in Figures 7(b1-b2) and Figures 7(c1-c2). As we can see, power reduction opportunity arises when the low memory referencing process starts to execute. Immediate Power Down (IPD) can significantly reduce power dissipated in Standby mode, whereas Immediate Self Refresh (ISR) can achieve additional energy benefit by also powering down the PLL and the registers, although at a severe performance penalty. We will look at their performance implications in detail shortly.

Table 2. Summary of the synthetic benchmark. All cycles are in unit of processor cycles.

		No Power Management	Immediate Power Down	Immediate Self Refresh	HW-only	HW–SW
Total Simulated Cycles	3,442,155,784 cycles					
Number of Read	10,906,196					
Number of Writes	11,055					
Energy Consumption		10.34 J	7.83 J	6.04 J	7.35 J	6.18 J
Average Power		6.01 W	4.55 W	3.51 W	4.27 W	3.59 W
Average Response Time		96.92 cycles	105.04 cycles	894.01 cycles	107.20 cycles	106.81 cycles
Delayed Accesses Due to PD		0	10,486,433	0	10,391,535	10,531,756
Delayed Accesses Due to SR		0	0	603,389	16,340	8,044

Next, Figure 7(d1) shows the results when power management decisions is dynamically made by the hardware (e.g., PMU in the memory controller). We assume IPD is implemented in the memory controller by default as it has a significant energy benefit and with only a very small performance impact (shown later). The PMU keeps history information on past accesses in its internal registers which are used to dynamically predict threshold values to determine after how long of an idle period before Self Refresh mode should be entered. It uses a moving window size of 500 μsec, which is reasonable because it can avoid over-compensation and provide good adaptability to realistic workloads. However, the result shows that it only outperforms the IPD strategy by approximately 6% in power because when the hardware tries to make power management decisions based on its observation on the past memory access behavior, it gets confused when two processes with very different access behaviors are accessing the same rank in an interleaved-manner. One can argue that if the window size is reduced to 100 μsec or even 10 μsec, we can adapt more quickly. However, shrinking the window size is a double-edged sword, having this better adaptability runs at a higher chance to over-aggressively predict threshold values from observing transient behaviors at run-time. Shrinking the window size can benefit this synthetic workload, but for realistic workloads, it can cause more harm than benefits. As we will show in the next section, mistakingly entering Self Refresh mode can be very expensive. Furthermore, as more and more systems are switching to smaller scheduling quanta (e.g., from 10 msec to 1 msec or even smaller) to increase responsiveness in the system, higher context switching rate will make the hardware predictor's job even more difficult.

Finally, in Figure 7(e1) we show that if the system software can inform the PMU in the memory controller of which process is currently running, more aggressive and accurate power management decisions can be made. The PMU used here is exactly the same as that described above, but with additional capabilities to keep the past access history specific to each process and to save/restore the history-keeping registers at each context switch. In this figure, we can see that immediately after the low memory referencing process starts to run, the PMU is able to instantaneously put the rank to Self Refresh, thus saving more energy. Additionally, unlike in the case of the HW-only technique, the cooperative technique will not be affected when the task scheduling quantum becomes increasingly smaller over time.

Results. The effect on energy can be easily obtained in our simulator. Performance implication is more difficult to quantify though, as it is limited by the trace-driven nature of this study. From a memory trace, we can identify exactly which memory reference is delayed and by how long due to power management. However, the dependency information among memory requests is not retained in a trace-based approach. Therefore, there is no way for us to know whether a delayed memory transaction will also delay a memory request that goes to an independent rank. To measure performance implication, instead, we use the average response time (service time) for each memory reference. This is shown in Table 2. In this table, we also summarized all other results for the synthetic workload.

From this table, we can see that using IPD is clearly beneficial. Compared to no power management, which has an average response time of 96.92 cycles per memory reference and consumes 10.34 J, IPD has an average response time of 105.04 cycles (+8.4%) and consumes only 7.83 J (-24.3%). A few percent increase in the average response time is usually not a big problem for server-type workloads as most are typically bandwidth-limited. On the other hand, when using Immediate Self Refresh, even though we can get an additional energy benefit (6.04 J, -41.6%), but it comes at a prohibitively high average response time (894.01 cycles, +822.42%). Compared to these static techniques, the dynamic ones perform much better. They consume almost as little energy as ISR but without ISR's hefty performance penalty, and they consume much less energy than IPD but pays almost as little performance penalty as IPD. Among the dynamic techniques, the HW–SW cooperative technique shows clear energy benefits over the HW-only approach. Specifically, it consumes 15.9% less energy than HW-only and also has a slightly better average response time. In Table 2, we also show the number of delayed requests due to exiting Power Down (PD) and Self Refresh (SR). Exiting PD is only 1 memory clock cycle, whereas exiting SR is much more expensive — 200 memory clock cycles. One of the reasons why HW–SW consumes less energy and has lower response time than the HW-only approach is that it can more accurately predict threshold for entering Self Refresh, and this is apparent from observing that HW–SW has far fewer number of delayed requests due to exiting from SR.

4.3 SPEC Benchmarks

One of the benchmarks we used in our evaluation is SPECjbb2000 [17]. It is implemented as a Java program emulating a 3-tier server system with an emphasis on the middle tier. The tiers simulate a typical business application, where users in Tier 1 generate inputs that result in the execution of business logic in the middle tier (Tier 2), which calls a database on the third tier. In a benchmark run, one can instantiate multiple warehouses, each with a 3-tier system. Each warehouse executes as a separate Java thread within the JVM, and is mapped to a different Linux process. However, since all warehouses are essentially running the same type of workload and they all share the same memory address space within the JVM, we will only observe a small amount of variation in how memory is accessed between context switches among these SPECjbb processes. In such systems, the benefit of using the HW–SW power management technique is limited. However, in real server systems, where the processor time is shared among multiple users and their applications, multiple server processes, and various daemon processes, we can expect memory access behavior to change constantly when context switching between these processes at a fine granularity. To emulate such a system, we decided to run a few SPECcpu2000 benchmarks with well known execution

Table 3. Summary of low memory-intensive workload

	No Power Management	IPD	ISR	SW-only (PAVM)	HW-only	HW–SW
Energy Consumption	353.11 J	194.04 J	56.73 J	114.92 J	105.33 J	87.79 J
Average Power	53.24 W	29.81 W	8.71 W	17.65 W	16.18 W	13.48 W
Average Response Time	114.77 cycles	126.06 cycles	1121.73 cycles	128.74 cycles	128.96 cycles	130.47 cycles
Delayed Accesses Due to PD	0	6,790,058	0	5,871,024	6,771,680	5,863,568
Delayed Accesses Due to SR	0	0	2,704,257	1,155	10,111	4,925

Table 4. Summary of high memory-intensive workload

	No Power Management	IPD	ISR	SW-only (PAVM)	HW-only	HW–SW
Energy Consumption	390.18 J	225.21 J	105.31 J	130.25 J	129.40 J	111.11 J
Average Power	56.42 W	32.56 W	15.23 W	18.83 W	18.71 W	16.07 W
Average Response Time	134.08 cycles	144.11 cycles	909.73 cycles	144.45 cycles	145.94 cycles	148.64 cycles
Delayed Accesses Due to PD	0	20,688,949	0	16,242.587	20,641,111	16,228,517
Delayed Accesses Due to SR	0	0	4,944,750	476	17,357	5,999

behavior in parallel with the SPECjbb workload. We classified these workloads as either *"high memory-intensive"* or *"low memory-intensive"*, based on L2 miss rates [3]. For the low memory-intensive workload, we run SPECjbb having 8 warehouses in parallel with *256.bzip2* and *186.crafty*, and for the high memory-intensive workload, we run SPECjbb in parallel with *181.mcf* and *179.art*. *Reference* input sets are used for these SPECcpu2000 benchmarks.

Results. In Table 3 and Table 4, we total energy, average power and average response time for the low memory-intensive and high memory-intensive workloads, respectively, for various power management techniques. IPD is assumed to be implemented in the memory controller to complement all other power management techniques (except for ISR) that we will evaluate. Here, we compare five techniques against each other — IPD, ISR, SW-only (PAVM), HW-only, and HW–SW.

First we look at the static techniques. IPD by itself uses much more power than the other techniques, and it has only a slightly better average response time than SW-only, HW-only, and HW–SW approaches, and therefore, is not useful by itself. ISR's prohibitively-high average responsive time makes it not practical to use either by itself. Dynamic techniques perform much better than these static techniques. Among the three dynamic techniques, PAVM performs the worst, and HW–SW performs the best in terms of power savings. For the low memory-intensive workload, HW–SW consumes 16.7% less energy than HW-only, and 23.6% less energy than SW-only. It also has a comparable average response time (130.47 cycles) to SW-only (128.74 cycles) and HW-only (128.96 cycles). For the high memory-intensive workload, HW–SW consumes 14.1% less energy than HW-only, and 14.7% less energy than SW-only, and it has only a slightly higher response time (148.64 cycles) than both SW-only (144.45 cycles) and HW-only (145.94 cycles) approaches.

5   Related Work

Recent research has demonstrated that a significant amount of energy can be saved in computing systems by exploiting power management capabilities built into modern hardware components. Among power management techniques for main memory, there are two main types of approaches — hardware and software-controlled. Among the hardware-controlled approaches, Lebeck *et al.* [2,7] studied the effects of various static and dynamic memory controller policies to reduce power with extensive simulation in a single-process environment. In another paper [8], they used stochastic Petri Nets to explore more complex policies. Delaluz *et al.* took a similar approach in [5], where they studied various flavors of threshold predictors and evaluated their energy implications. The techniques proposed in this paper are orthogonal to the works described above and can be used to improve the prediction accuracy in some of these previously-proposed threshold prediction mechanisms. However, unlike these previous works, the techniques proposed in this paper are specifically designed and optimized for a multitasking environment, as are most of today's systems. Furthermore, we have also taken into account of various OS effects, which were shown to be also important in practice [9].

Among the software-controlled approaches, Delaluz *et al.*[6] demonstrated a simple scheduler-based power management policy. Huang *et al.* [9] later implemented Power-Aware Virtual Memory (PAVM) to improve upon this work. PAVM modifies the underlying physical page allocator to make it more energy-efficient by collaborating with the virtual memory through a NUMA management layer so that the energy footprint of each process is reduced. To cope with various dynamics in real systems, PAVM leverages advanced techniques, such as library aggregation and page migration. Delaluz *et al.*[5] have also proposed a compiler-directed approach, where power management decisions are statically determined. Due to its static nature, this approach is not very appropriate for most complex systems, but may be applicable in some embedded systems where workloads are more deterministic.

There are advantages and disadvantages in the two types of approaches. The cooperative technique that we proposed in this paper offers the best features in both. With minimal help from the system software, we are able to show that the PMU in the memory controller can more accurately monitor memory traffic and thus more efficiently managing power. In other research contexts, using software and hardware collaboration [1,12] has also been shown to be beneficial in terms of improving performance and security, and providing new functionalities.

6   Conclusion

In this paper, we proposed a novel power management technique that makes use of co-operation between the system software and the memory controller hardware. It is shown to make a significant improvement in the accuracy of the PMU's threshold prediction logic. Using a full-system simulator, our HW–SW cooperative approach is shown to consume 14.2–17.3% less energy than the HW-only technique and 16.0–25.8% less energy than the SW-only technique. We used a uni-processor system to explore the feasibility of using this technique and quantified its benefits. We are planning to extend this work to multi-processor systems, where a combination of running processes, instead of a single running process, must be considered.

Furthermore, alternative to this software-assisted hardware power management technique proposed here, we can also imagine scenarios where the hardware can also provide feedback to the system software to create additional energy saving opportunities. For example, the hardware can inform the OS how "hot" each of the physical pages are being accessed, and the OS can use this information to re-arrange memory pages within each process's address space. This allows us to either (1) run hot ranks hotter and cold ranks colder to create more energy saving opportunities in the cold ranks, or (2) balance power dissipation on each rank and remove hot spots. Additionally, we would also like to explore direct cooperation between applications and the PMU. As applications themselves know more about their future memory access behavior than the OS, such information can prove to be beneficial to the memory controller in its prediction logic, and thus, can be used to further enhance the proposed power management system.

References

1. D. Engler, M. Kaashoek, and J. O'Toole Jr. Exokernel: An operating system architecture for application-level resource management. In *15th ACM Symposium on Operating Systems Principles (SOSP)*, 1995.
2. A. R. Lebeck *et al.* Power aware page allocation. In *Architectural Support for Programming Languages and Operating Systems (ASPLOS)*, pages 105–116, 2000.
3. Karthikeyan Sankaralingam *et. al.* Exploiting ilp, tlp, and dlp with the polymorphous trips architecture. In *ISCA*, 2003.
4. P. Bohrer *et al.* Mambo — a full system simulator for the powerpc archtecture. In *ACM SIGMETRICS Performance Evaluation Review*, volume 31, 2004.
5. V. Delaluz *et al.* DRAM energy management using software and hardware directed power mode control. In *International Symposium on High-Performance Computer Architecture*, pages 159–170, 2001.
6. V. Delaluz *et al.* Scheduler-based DRAM energy power management. In *Design Automation Conference 39*, pages 697–702, 2002.
7. X. Fan, C. S. Ellis, and A. R. Lebeck. Memory controller policies for DRAM power management. In *International Symposium on Low Power Electronics and Design (ISLPED)*, pages 129–134, 2001.
8. X. Fan, C. S. Ellis, and A. R. Lebeck. Modeling of DRAM power control policies using deterministic and stochastic petri nets. In *Workshop on Power-Aware Computer Systems*, 2002.
9. H. Huang, P. Pillai, and K. G. Shin. Design and implementation of power-aware virtual memory. In *USENIX Annual Technical Conference*, pages 57–70, 2003.
10. Y. Joo and *el al.* Energy exploration and reduction of SDRAM memory systems. In *DAC*, pages 892–897, 2003.
11. C. Lefurgy, K. Rajamani, F. Rawson, W. Felter, M. Kistler, and Tom Keller. Energy management for commercial servers. In *IEEE Computer*, pages 39–48, Dec 2003.
12. David Lie, Chandramohan A. Thekkath, and Mark Horowitz. Implementing an untrusted operating system on trusted hardware. In *19th ACM Symposium on Operating Systems Principles (SOSP)*, 2003.
13. Mesquite Software. http://www.mesquite.com.
14. Micron. http://download.micron.com/pdf/technotes/tn4603.pdf.

15. Micron. http://www.micron.com.
16. H. Shafi, P. J. Bohrer, J. Phelan, C. A. Rusu, and J. L. Peterson. Design and validation of a performance and power simulator for PowerPC systems. In *IBM Journal on Research and Development*, volume 47, 2003.
17. Standard Performance Evaluation Corporation (SPEC). http://www.specbench.org/jbb2000/.
18. Standard Performance Evaluation Corporation (SPEC).   http://www.specbench.org/osg/cpu2000/.
19. W. Wulf and S. McKee. Hitting the memory wall: Implications of the obvious. *Computer Architecture News*, 23(1):20–24, 1995.

Energy-Aware Data Prefetching
for General-Purpose Programs

Yao Guo[1], Saurabh Chheda[2], Israel Koren[1], C. Mani Krishna[1],
and Csaba Andras Moritz[1]

[1] ECE Dept., University of Massachusetts, Amherst, MA 01003
{yaoguo, koren, krishna, andras}@ecs.umass.edu
[2] BlueRISC Inc., Hadley, MA 01035
chheda@bluerisc.com

Abstract. There has been intensive research on data prefetching focusing on performance improvement, however, the energy aspect of prefetching is relatively unknown. Our experiments show that although software prefetching tends to be more energy efficient, hardware prefetching outperforms software prefetching on most of the applications in terms of performance. This paper proposes several techniques to make hardware-based data prefetching power-aware. Our proposed techniques include three compiler-based approaches which make the prefetch predictor more power efficient. The compiler identifies the pattern of memory accesses in order to selectively apply different prefetching schemes depending on predicted access patterns and to filter out unnecessary prefetches. We also propose a hardware-based filtering technique to further reduce the energy overhead due to prefetching in the L1 cache. Our experiments show that the proposed techniques reduce the prefetching-related energy overhead by close to 40% without reducing its performance benefits.

1 Introduction

In recent years, energy and power efficiency have become key design objectives in microprocessors, in both embedded and general-purpose domains. Although considerable research [18,3,16,17,6,13,10,11] has been focused on improving the performance of prefetching mechanisms, the impact of prefetching techniques on processor energy efficiency has not yet been fully investigated.

Our experiments [8] on five hardware-based data prefetching techniques show that while aggressive prefetching techniques often help to improve performance, in most of the applications, they increase memory system energy consumption by as much as 30%. In many systems [7,12], this is equivalent to more than 15% increase in chip-wide energy consumption.

We implemented two software prefetching techniques [14,11] to compare the performance and energy efficiency of hardware and software prefetching. The results show that in general software prefetching is more energy-efficient while hardware prefetching yields better performance for most applications. In this paper, we focus on making one of the hardware prefetching techniques (which

B. Falsafi and T.N. Vijaykumar (Eds.): PACS 2004, LNCS 3471, pp. 78–94, 2005.
© Springer-Verlag Berlin Heidelberg 2005

yields the best performance speedup) more energy-efficient without sacrificing its performance benefits.

Aggressive hardware prefetching is beneficial in many applications as it helps to hide memory-system related performance costs. By doing that, however, it often significantly increases energy consumption in the memory system. The memory system consumes a large fraction of the total chip-energy and it is therefore a key area targeted for energy optimizations. Our experiments show that most of the energy degradation is due to the prefetch-hardware related energy costs and unnecessary L1 data-cache lookups related to prefetches that hit in the L1 cache.

We propose several power-aware techniques for hardware data prefetching to reduce the energy overheads stated above. The techniques include:

- A compiler-based prefetch filtering approach, which reduces energy consumption by only searching the prefetch hardware tables for selective memory instructions identified by the compiler;
- A compiler-assisted selective prefetching mechanism, which utilizes compiler supplied static information to selectively apply different prefetching schemes depending on predicted access patterns;
- A compiler-driven filtering technique using a runtime stride counter designed to reduce prefetching energy consumption wasted on memory access patterns with very small strides; and
- A hardware-based filtering technique applied to further reduce the L1 cache related energy overhead due to prefetching.

The SimpleScalar [5] simulation tool has been modified to implement the different prefetching techniques and collect statistics on performance as well as switching activity in the memory system. The compiler passes for both software prefetching and power-aware hardware prefetching are implemented using the SUIF infrastructure [19]. To estimate power consumption in the memory system, we use state-of-the-art low-power cache circuits and simulate them using HSpice. Our experiments show that the proposed techniques successfully reduce the prefetching-related energy overheads by 40% on average, without reducing the performance benefits of prefetching.

The rest of this paper is organized as follows. Section 2 describes the energy overhead of data prefetching. The energy-aware prefetching solutions are presented in Sec. 3. Section 4 presents the experimental assumptions. Section 5 gives a detailed analysis of the results. We conclude with Sec. 6.

2 Motivation

Our previous work [8] has evaluated the energy perspective of hardware-based data prefetching techniques. In addition to the hardware techniques evaluated, we also implemented two software prefetching techniques [14,11] and compare their performance and energy consumptions to the hardware mechanisms. More details and background on the this section can be found in [8].

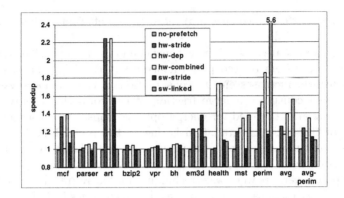

Fig. 1. Performance speedup for different prefetching schemes

To explore the energy aspects of data prefetching techniques, we provide experimental results for the following five prefetching techniques:

– *Stride prefetching* [3] - Focuses on array-like structures, it catches constant strides in memory accesses and prefetches using the stride information;
– *Dependence-based prefetching* [16] - Designed to prefetch on pointer-intensive programs containing linked data structures where no constant strides can be found;
– A *combined* stride and dependence-based approach - Focuses on general-purpose programs, which often use both array and pointer structures, to achieve benefits from both stride and pointer prefetching.
– *Compiler-based prefetching* similar to [14] - Use the compiler to insert prefetch instructions for strided array accesses.
– *Compiler-based prefetching on Linked Data Structures* - Uses the greedy approach in [11] to prefetch pointer structures.

The first three techniques are hardware-based and they require the help of one or more hardware history tables to trigger prefetches. The last two are software-based techniques which use compiler analysis to decide what addresses should be prefetched and where in the program to insert the prefetch instructions.

The performance improvement of the five prefetching techniques is shown in Fig. 1. The first five benchmarks are from SPEC2000 benchmarks; the last five are Olden benchmarks which contains many pointers and linked data structures.

As we expected, stride prefetching does very well on performance for SPEC2000 benchmarks, averaging just over 25% speedup across the five applications studied. In contrast, the dependence-based approach achieves an average speedup of 27% on the five Olden benchmarks. The combined approach achieves the best performance speedup among the three hardware techniques, averaging about 40%. In general, the combined technique is the most effective approach for general-purpose programs (which typically contain both array and pointer structures).

For the two software techniques, the compiler-based technique for strided accesses achieves almost 60% speedup on *art* and about 40% on *em3d*, with an

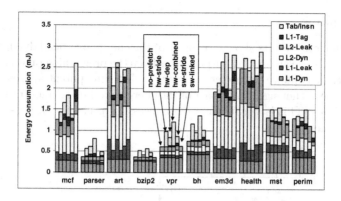

Fig. 2. Total cache energy consumption

average of 16% in performance speedup. The scheme for linked data structures yields an average of 55%, but it does extremely well on *perim*(a speedup of 5.6x). Without *perim*, the average speedup goes down to just 10%.

We calculated the total energy consumption in the memory system for each prefetching technique based on HSpice; for more details, see Sec. 4. The results are shown in Fig. 2. In the figure, we show the energy breakdown for (from bottom to top for each bar) L1 dynamic energy, L1 leakage, L2 dynamic energy, L2 leakage, L1 tag lookups due to prefetching, and prefetch hardware table accesses for hardware prefetching or prefetch instruction overhead for software prefetching.

The results in Fig. 2 show that the three hardware-based prefetching schemes result in a significant energy consumption overhead, especially in the combined prefetching approach. The average increase for the combined approach is more than 28%, which is mainly due to the prefetch table accesses and the extra L1 tag lookups due to prefetching. Software prefetching also increases energy consumption for most of the benchmarks, especially in *mcf* and *em3d*. However, compared to the combined hardware prefetching, software prefetching techniques are more energy-efficient for most of the benchmarks.

Considering both performance and energy-efficiency, it seems that there is no single prefetching solution which would yield the best performance and at the same time consume the least energy consumption. Based on our observation, the combined hardware-based technique outperforms others in terms of speedup for most benchmarks although it consumes considerably more energy than the other four techniques. The question is: can we make the combined hardware prefetching more energy-efficient without sacrificing its performance benefits?

3   Energy-Aware Prefetching Techniques

In this section, we will discuss how to reduce the energy overhead for the most aggressive hardware prefetching scheme, the combined stride and pointer prefetching. This scheme gives the best performance speedup for general-purpose programs, but it is the worst in terms of energy efficiency.

3.1   Overview

Our experimental results show that most of the energy overhead due to prefetching comes from two areas. The major part is from the prefetching prediction phase: when we search/update the prefetch history table to find potential prefetching opportunities; Another significant part of the energy overhead comes from the extra L1 tag-lookups. This is because many unnecessary prefetches are issued by the prefetch engine.

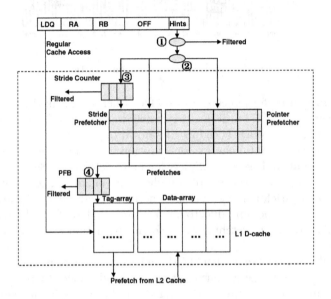

Fig. 3. Power-aware prefetching architecture for general-purpose programs

Figure 3 shows the modified combined prefetching architecture including four energy-saving components. The first three techniques are compiler-based approaches used to reduce prefetch-table related costs and some extra L1 tag lookups due to prefetching. The last one is a hardware-based approach designed to reduce the extra L1 tag lookups. The techniques proposed, as numbered in Fig. 3, are:

1. A compiler-based prefetch filtering approach which reduces prefetch hardware energy cost by only searching the prefetch hardware tables for memory instructions selected by the compiler;
2. A compiler-assisted selective prefetching mechanism which utilizes the compiler supplied static information to selectively apply different prefetching schemes depending on predicted access patterns;
3. A compiler-driven filtering technique using a runtime stride counter, designed to reduce prefetching attempts and energy consumption wasted on memory access patterns with very small strides; and

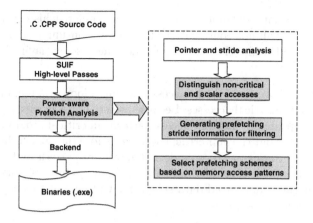

Fig. 4. Compiler analysis used for power-aware prefetching

4. A hardware-based filtering technique applied to further reduce the L1 cache-related energy overhead due to prefetching.

The compiler-based approaches help make the prefetch predictor more selective based on program information extracted. With the help of the compiler hints, the energy-aware prefetch engine performs much fewer searches in the prefetch hardware tables and issues fewer prefetches, which results in less energy overhead being consumed in L1 cache tag-lookups.

Figure 4 shows the compiler passes in our approach. Prefetch analysis is the process where we generate the prefetching hints, including whether or not we will do prefetching, which prefetcher to choose, and the stride information. A speculative pointer and stride analysis approach [9] is applied to help analyze the programs and generate the information we need for prefetch analysis. Compiler-assisted techniques require the modification of the instruction set architecture to encode the prefetch hints generated by the compiler analysis. These hints could be accommodated by reducing the number of offset bits. We will discuss how to perform the analysis for each of the techniques in detail later.

In addition, our hardware-based filtering technique utilizes the temporal and spatial locality of prefetching requests to filter out the requests trying to prefetch the same cache line as prefetched recently. The technique is based on a small hardware buffer called the Prefetch Filtering Buffer (PFB).

3.2   Compiler-Based Prefetch Filtering (CBPF)

One of our observations is that not all load instructions are useful for prefetching. Some instructions, such as scalar memory accesses, have no access patterns and cannot anyway trigger useful prefetches when fed into the prefetcher.

We use the compiler to distinguish memory accesses useful for prefetching from those which my have no benefit. Only those useful load instructions, selected by the compiler, are fed into the prefetcher. Instructions identified with "no prefetching potential" will not be added to the prefetch history table. Thus, these instructions will not contribute to the energy consumption overhead.

The compiler identifies the following memory accesses as having "no prefetching potential":

- *Non-critical accesses*: Memory accesses within a loop or a recursive function are regarded as critical accesses. Because prefetching schemes are anyway designed to capture the memory access patterns in critical program phases, we can safely filter out the non-critical accesses before they reach the prefetcher.
- *Scalar accesses*: Scalar accesses do not have any pattern and will not contribute to the prefetcher if fed into the prefetcher. Only memory accesses to array structures and linked data structures will be sent to the prefetcher to make prefetching decisions.

The instructions selected by the compiler are annotated with "no prefetching potential" and are filtered out before they are fed into the prefetcher. This optimization could eliminate on average as much as 8% of all the prefetch table accesses, as we will show later.

3.3   Compiler-Assisted Selective Prefetching (CASP)

Another compiler approach focuses on how to help the prefetch predictor to choose one of the prefetching schemes in the combined prefetching approach.

One important aspect of the combined approach is that it uses two techniques independently and prefetches based on the memory access patterns for all memory accesses. As we know, stride prefetching works better on array-based accesses and dependence-based prefetching is more appropriate for pointer-based structures. One obvious approach is therefore to distinguish these two types of accesses.

Distinguishing between pointers and non-pointer accesses is difficult during execution time. However, we can distinguish them easily during compilation passes. Array accesses and pointer accesses are annotated using hints written into the instructions. During runtime, the prefetch engine can identify the hints and apply different prefetching mechanisms.

We have found that simply splitting the array and pointer structures is not very effective and affects the performance speedup (which is the primary goal of prefetching techniques). Instead, we use the following heuristic to decide whether we should use stride prefetching or pointer prefetching:

- Memory accesses to an array which does not belong to any larger structure (e.g., fields in a C struct) are only fed into the stride prefetcher;
- Memory accesses to an array which belongs to a larger structure are fed into both stride and pointer prefetchers;
- Memory accesses to a linked data structure with no arrays are only fed into the pointer prefetcher;
- Memory accesses to a linked data structure that contains arrays are fed into both prefetchers.

The above heuristic is able to preserve the performance speedup benefits of the aggressive prefetching scheme. We can filter out up to 20% of all the prefetch-table accesses and up to 10% of the extra L1 tag lookups due to prefetching, by applying this technique.

3.4 Compiler-Hinted Filtering Using a Runtime Stride Counter (SC)

Another part of prefetching energy overhead comes from memory accesses with small strides. Accesses with very small strides (compared to the cache line size of 32 bytes we use) could result in frequent accesses to the prefetch table and issuing more prefetch requests than needed. For example, if we have an iteration on an array with a stride of 4 bytes, we will access the hardware table at least 8 times before we reach the point where we can issue a useful prefetch to get a new cache line. The overhead not only comes from the extra prefetch table accesses; 8 different prefetch requests are also issued to prefetch the same cache line during the 8 iterations.

Software prefetching would be able to avoid the penalty by doing loop unrolling. In our approach, we use hardware to accomplish loop unrolling with assistance from the compiler. The compiler predicts as many strides as possible based on static information. Stride analysis is applied not only for array-based memory accesses, but we also predict strides for pointer accesses with the help of pointer analysis. Detailed information on how to do the pointer and stride analysis could be found in [9].

Strides predicted as larger than half of the cache line size (16 bytes) will be considered as large enough since they will be able to reach a different cache line after each iteration. Strides smaller than the half of the cache line size will be recorded and passed to the hardware. This is a very small 8-entry buffer used to record the most recently used instructions with small strides. Each entry contains the program counter (PC) of the particular instruction and a stride counter. The counter is used to count how many times the instruction occurs after it was last fed into the prefetcher. The counter will be set to a maximal value (decided by cache_line_size/stride) and is then decremented by one each time the instruction is executed. The instruction is only fed into the prefetcher when its counter is decreased to zero; then, the counter will be reset to the maximum value.

For example, if we have an array access (in a loop) with a stride of 4 bytes, the counter will be set to 8 initially. Thus, during eight occurrences of this load instruction, only once it is sent to the prefetcher.

This technique reduces 5% of all the prefetch table accesses as well as 10% of the extra L1 cache tag lookups, while resulting in less than 0.3% performance degradation.

3.5 Hardware Prefetch Filtering Using PFB

To further reduce the L1 tag-lookup related energy consumption, we add a hardware-based prefetch filtering technique. Our approach is based on a very small hardware buffer called the Prefetch Filtering Buffer(PFB).

When a prefetch engine predicts a prefetching address, it does not prefetch the data from that address immediately from the lower-level memory system (e.g., L2 Cache). Typically, tag lookups on L1 tag-arrays are performed. If the data to be prefetched already exists in the L1 Cache, the prefetch request from the prefetch engine is dropped. A cache tag-lookup costs much less energy compared to a full read/write access to the low-level memory system (e.g., the L2 cache). However, associative tag-lookups are still energy expensive.

To reduce the number of L1 tag-checks due to prefetching, we add a PFB to remember the most recently prefetched cache tags. We check the prefetching address against the PFB when a prefetching request is issued by the prefetch engine. If the address is found in the PFB, the prefetching request is dropped and we assume that the data is already in the L1 cache. When the data is not found in the PFB, we perform normal tag lookup and proceed according to the lookup results. The LRU replacement algorithm is used when the PFB is full. The prefetch filtering scheme using the PFB is shown in Fig. 3.

A smaller PFB costs less energy per access, but can only filter out a smaller number of useless prefetches. A larger PFB can filter out more useless prefetches, but each access to the PFB costs more energy. To find out the optimal size of the PFB, we simulated a set of benchmarks with PFB sizes of 1 to 16. We will show in Sec. 5 that an 8-entry PFB is large enough to accomplish the prefetch filtering task with very small performance overhead.

PFBs are not always correct in predicting whether the data is still in L1 since the data might have been replaced although its address is still present in the PFB. We call this case a PFB misprediction. High PFB mispredictions would result in performance loss because useful prefetches are dropped. Fortunately, as we will show later, the PFB misprediction rate is very low (close to 0).

4   Experimental Assumptions

4.1   Experimental Framework

We implement the hardware-based data prefetching techniques by modifying the SimpleScalar [5] simulator. The software prefetching schemes are implemented using SUIF [19] and simulated with the modified SimpleScalar which can recognize prefetch instructions. We also use SUIF to implement the compiler passes for power-aware prefetching, generating annotations for all the prefetching hints which we later transfer to assembly codes. The binaries input to the SimpleScalar simulator are created using a native Alpha assembler. The parameters we use for the simulations are listed in Table 1.

The benchmarks evaluated are listed in Table 2. The SPEC2000 benchmarks [1] use mostly array-based data structures, while the Olden benchmark suite [15] contains pointer-intensive programs that make substantial use of linked data structures. We randomly select a total of ten benchmark applications, five from SPEC2000 and five from Olden. For SPEC2000 benchmarks, we fast forward the first one billion instructions and then simulate the next 100 million

Table 1. Baseline parameters

Processor speed	1GHz
Issue	4-way, out-of-order
L1 D-cache	32KB, CAM-tag, 32-way, 32bytes cache line
L1 I-cache	32KB, 2-way, 32bytes cache line
L1 cache latency	1 cycle
L2 cache	unified, 256KB, 4-way, 64bytes cache line
L2 cache latency	12 cycle
Memory latency	100 cycles latency + 10 cycles/word

Table 2. SPEC2000 & Olden benchmarks

Benchmark	Description
SPEC2000	
181.mcf	Combinatorial Optimization
197.parser	Word Processing
179.art	Image Recognition / Neural Nets
256.bzip2	Compression
175.vpr	Versatile Place and Route
Olden	
bh	Barnes & Hut N-body Algorithm
em3d	Electromagnetic Wave Propagation
health	Colombian Health-Care Simulation
mst	Minimum Spanning Tree
perimeter	Perimeters of Regions in Images

instructions. The Olden benchmarks are simulated to completion except for one (perimeter), since they complete in relatively short time.

4.2 Energy Modeling

To accurately estimate power and energy consumption in the L1 and L2 caches, we perform circuit-level simulations using HSpice. We base our design on a recently proposed low-power circuit [20] that we implemented in 100-nm BPTM technology. Our L1 cache includes the following low-power features: low-swing bitlines, local word-line, CAM-based tags, separate search lines, and a banked architecture. The L2 cache we evaluate is based on a banked RAM-tag design.

As we expect that implementations in 100-nm technology would have significant leakage, we apply a recently proposed circuit-level leakage reduction technique called asymmetric SRAM cells [2]. This is necessary because otherwise our conclusions would be skewed due to very high leakage power. The *speed enhanced cell* in [2] has been shown to reduce L1 data cache leakage by 3.8X for SPEC2000 benchmarks with no impact on performance. For L2 caches, we use the *leakage enhanced cell* which increases the read time by 5%, but can reduce leakage power by at least 6X. In our evaluation, we assume speed-enhanced cells for L1 and leakage enhanced cells for L2 data caches, by applying the different asymmetric cell techniques respectively.

The power consumption for our L1 and L2 caches are shown in Table 3.

Table 3. Cache configuration and power consumption

Parameter	L1	L2
size	32KB	256KB
tag array	CAM-based	RAM-based
associativity	32-way	4-way
bank size	2KB	4KB
# of banks	16	64
cache line	32B	64B
Power (mW)		
tag	6.5	6.3
read	9.5	100.5
write	10.3	118.6
leakage	3.1	23.0
reduced leakage	0.8	1.5

If an L1 miss occurs, energy is consumed not only in L1 tag-lookup, but also when writing the requested data back to L1. L2 accesses are similar, except that an L2 miss goes to off-chip main memory.

Table 4. Prefetch hardware table and power consumption

Table implementation	64×64 CAM-array
P_update (including lookup)	7.4mW
P_lookup	7.3mW

Each prefetching history table is implemented as a 64×64 fully-associated CAM-array. The power consumption for each lookup is 7.3mW and each update to the table costs 7.4mW based on HSpice simulation. The power numbers are shown in Table 4. The leakage energy of these hardware tables are very small compared to L1 and L2 caches due to their small area.

For software prefetching, the cost of the execution of a prefetch instruction includes an access to the L1 instruction cache by the prefetch instruction, and the pipeline cost of instruction fetching, decoding, and the calculation of prefetching addresses. These extra costs will increase the total energy consumption. Each L1 instruction cache access consumes about the same energy as an L1 data cache access, and the rest of the execution costs is generally comparable to an L1 data cache access [4]. Thus we assume that each prefetch instruction executed would consume an extra cost of roughly two times the L1 cache read energy cost in Fig. 2.

5   Results and Analysis

We simulated each of the four energy-saving techniques and evaluated their impact on energy consumption as well as performance speedup. All the techniques are applied to the combined stride and dependence-based pointer prefetching. We first show the results by applying each of the four techniques individually; and then, we apply them together in order.

5.1   Compiler-Based Techniques

Figure 5 shows the results for the three compiler-based techniques, first separately and then combined. The results shown are normalized to the baseline, which is the combined stride and pointer prefetching scheme without any of the new techniques.

Figure 5(a) shows the number of prefetch table accesses. The compiler-based prefetching filtering (CBPF) works best for *parser*: more than 33% of all the prefetch table accesses are eliminated. On average, CBPF achieves about 7% reduction in prefetch table accesses. The compiler-assisted selective prefetching (CASP) achieves the best reduction for *health*, about 20%, and on average saves 6%. The stride counter filtering (SC) technique removes 12% of prefetch table accesses for *bh*, with an average of over 5%. The three techniques combined filter out more than 20% of the prefetch table accesses for five of the ten benchmarks, with an average of 18% across all applications.

Figure 5(b) shows the extra L1 tag lookups due to prefetching. CBPF reduces the tag lookups by more than 8% on average; SC removes about 9%. CASP does not show a lot of savings, averaging just over 4%. The three techniques combined achieve tag-lookup savings of up to 35% for *bzip2*, averaging 21% compared to the combined prefetching baseline.

The performance penalty introduced by the three techniques is shown in Fig. 5(c). As shown, the performance impact is negligible. The only exception is *em3d*, which has less than 3% of performance degradation, due to filtering using SC.

5.2   Prefetch Filtering Using PFB

Prefetch filtering using PFB will filter out those prefetch requests which would result in a L1 cache hit if issued. We simulated different sizes of PFB to find out the best PFB size, considering both performance and energy consumption.

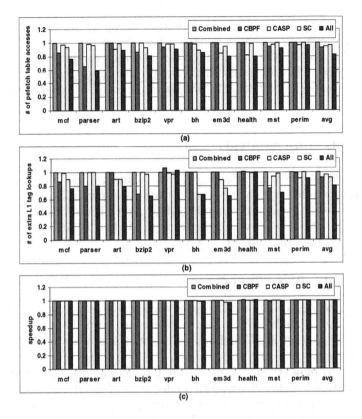

Fig. 5. Simulation results for the three compiler-based techniques: (a) normalized number of the prefetch table accesses; (b) normalized number of the L1 tag lookups due to prefetching; and (c) impact on performance

Figure 6 shows the number of L1 tag lookups due to prefetching after applying the PFB prefetch filtering technique with PFB sizes ranging from 1 to 16.

As we can see from the figure, even a 1-entry PFB can filter out about 40% of all the prefetch tag accesses (on average). An 8-entry PFB can filter out over 70% of tag-checks with almost 100% accuracy. Increasing the PFB size to 16 does not increase the filtering percentage significantly. The increase is about 2% on the average compared to an 8-entry PFB, while the energy cost per access doubles.

We also show the ideal situation (OPT in the figure), where all the prefetch hits are filtered out. For some of the applications, such as *art* and *perim*, the 8-entry PFB is already very close to the optimal case. This shows that an 8-entry PFB is a good enough choice for this prefetch filtering.

As we stated before, PFB predictions are not always correct: it is possible that a prefetched address still resides in the PFB but it does not exist in the L1 cache (it has been replaced). The number of PFB mispredictions is shown in Table 5. Although the number of mispredictions increases with the size of

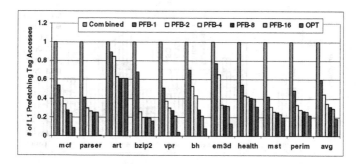

Fig. 6. The number of L1 tag lookups due to prefetching after applying the hardware-based prefetch filtering technique with different sizes of PFB

Table 5. The number of PFB mispredictions for different sizes of PFBs

Bench	PFB-1	PFB-2	PFB-4	PFB-8	PFB-16
mcf	0	0	0	1	9
parser	0	0	0	0	0
art	0	0	0	0	0
bzip2	0	0	0	0	0
vpr	0	0	0	0	0
bh	0	0	0	0	0
em3d	0	0	0	0	0
health	0	0	0	0	1
mst	0	0	11	11	11
perimeter	0	0	0	0	0

the PFB, an 8-entry PFB makes almost perfect predictions and does not affect performance.

5.3   Energy Savings

We apply the techniques in the following order CBPF, CASP, SC, and PFB. We show the energy savings after each technique is added in Fig. 7.

Compared to the combined stride and pointer prefetching, the compiler-based prefetch filtering (CBPF) shows good improvement for *mcf* and *parser*, with an average reduction of total memory system energy of about 3%.

The second scheme, compiler-assisted selective prefetching (CASP), reduces the energy consumed by about 2%, and shows good improvement for *health* and *em3d* (about 5%).

The stride counter approach is then applied. It reduces the energy consumption for both prefetch hardware tables and L1 prefetch tag accesses. It improves the energy consumption consistently for almost all benchmarks, achieving an average of just under 4% savings on the total energy consumption.

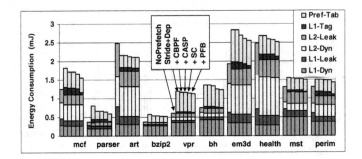

Fig. 7. Energy consumption in the memory system after applying different energy-aware prefetching schemes

Finally, the prefetch filtering technique is applied with an 8-entry PFB. The PFB reduces more than half of the L1 prefetch tag lookups and improves the total energy consumption by about 3%.

Overall, the four power-saving techniques together reduce by almost 40% the energy overhead of the combined prefetching approach: the energy overhead due to prefetching is reduced from 28% to 17%. This is about 11% of the total memory system energy (including L1, L2 caches and prefetch tables).

5.4 Performance Degradation

Figure 8 shows the performance statistics associated with each of the four techniques.

We can see that there is no performance impact except for *em3d* where stride-filtering yields less than 3% speedup degradation. On average, the performance degradation is only 0.4%, while we achieve an average energy saving of 11%.

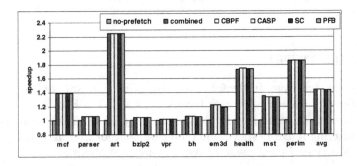

Fig. 8. Performance speedup after applying different energy-aware prefetching schemes

6   Conclusion

This paper explores the energy-efficiency aspects of data-prefetching techniques and proposes several new techniques to make prefetching energy-aware. Our proposed techniques include three compiler-based approaches which help to make the prefetch predictor more selective and filter out unnecessary prefetches based on static program information. We also propose a hardware based filtering technique to further reduce the energy overheads due to prefetching in the L1 cache. Our experiments show that the proposed techniques combined reduce the prefetching-related energy overheads by 40%, with almost no impact on performance.

References

1. The standard performance evaluation corporation, 2000. http://www.spec.org.
2. N. Azizi, A. Moshovos, and F. N. Najm. Low-leakage asymmetric-cell sram. In *Proc. of the 2002 international symposium on Low power electronics and design*, pages 48–51, 2002.
3. J. L. Baer and T. F. Chen. An effictive on-chip preloading scheme to reduce data access penalty. In *Supercomputing 1991*, pages 179–186, Nov. 1991.
4. D. Brooks, V. Tiwari, and M. Martonosi. Wattch: A framework for architectural-level power analysis and optimizations. In *Proceedings of the 27th Annual International Symposium on Computer Architecture*, pages 83–94, June 2000.
5. D. C. Burger and T. M. Austin. The simplescalar tool set, version 2.0. Technical Report CS-TR-1997-1342, University of Wisconsin, Madison, June 1997.
6. R. Cooksey, S. Jourdan, and D. Grunwald. A stateless content-directed data prefetching mechanism. In *Tenth intl. conf. on architectural support for programming languages and operating systems(ASPLOS-X)*, pages 279–290, 2002.
7. M. K. Gowan, L. L. Biro, and D. B. Jackson. Power considerations in the design of the alpha 21264 microprocessor. In *Proceedings of the 1998 Conference on Design Automation (DAC-98)*, pages 726–731, June 1998.
8. Y. Guo, S. Chheda, I. Koren, C. M. Krishna, and C. A. Moritz. Energy characterization of hardware-based data prefetching. In *International Conference on Computer Design (ICCD'04)*, Oct. 2004.
9. Y. Guo, S. Chheda, and C. A. Moritz. Runtime biased pointer reuse analysis and its application to energy efficiency. In *Workshop on Power-Aware Computer Systems(PACS) at Micro-36*, pages 1–15, Dec. 2003.
10. M. H. Lipasti, W. J. Schmidt, S. R. Kunkel, and R. R. Roediger. Spaid: software prefetching in pointer- and call-intensive environments. In *Proceedings of the 28th annual international symposium on Microarchitecture*, pages 231–236, Nov. 1995.
11. C.-K. Luk and T. C. Mowry. Compiler-based prefetching for recursive data structures. In *Architectural Support for Programming Languages and Operating Systems (ASPLOS-VII)*, pages 222–233, Oct. 1996.
12. J. Montanaro and et. al. A 160-MHz, 32-b, 0.5-W CMOS RISC microprocessor. *Digital Technical Journal of Digital Equipment Corporation*, 9(1), 1997.
13. T. Mowry. *Tolerating Latency Through Software Controlled Data Prefetching*. PhD thesis, Dept. of Computer Science, Stanford University, Mar. 1994.
14. T. C. Mowry, M. S. Lam, and A. Gupta. Design and evaluation of a compiler algorithm for prefetching. In *Fifth International Conference on Architectural Support for Programming Languages and Operating Systems*, pages 62–73, Oct. 1992.
15. A. Rogers, M. C. Carlisle, J. H. Reppy, and L. J. Hendren. Supporting dynamic data structures on distributed-memory machines. *ACM Transactions on Programming Languages and Systems*, 17(2):233–263, Mar. 1995.

16. A. Roth, A. Moshovos, and G. S. Sohi. Dependence based prefetching for linked data structures. In *Proc. of the 8th Architectural Support for Programming Languages and Operating Systems)*, pages 115–126, oct 1998.
17. A. Roth and G. S. Sohi. Effective jump-pointer prefetching for linked data structures. In *Proceedings of the 26th annual international symposium on Computer architecture*, pages 111–121. IEEE Computer Society Press, 1999.
18. A. J. Smith. Sequential program prefetching in memory bierarchies. *IEEE Computer*, 11(12):7–21, Dec. 1978.
19. R. Wilson, R. French, C. Wilson, S. Amarasinghe, J. Anderson, S. Tjiang, S.-W. Liao, C.-W. Tseng, M. W. Hall, M. Lam, and J. L. Hennessy. SUIF: A parallelizing and optimizing research compiler. Technical Report CSL-TR-94-620, Computer Systems Laboratory, Stanford University, May 1994.
20. M. Zhang and K. Asanovic. Highly-associative caches for low-power processors. In *Kool Chips Workshop, 33rd International Symposium on Microarchitecture*, Dec. 2000.

Bus Power Estimation and Power-Efficient Bus Arbitration for System-on-a-Chip Embedded Systems

Ke Ning[1,2] and David Kaeli[2]

[1] Department of Electrical and Computer Engineering, Northeastern University, Boston MA 02115
[2] Analog Devices Inc, 3 Technology Way, Norwood MA 02062

Abstract. In a system-on-a-chip embedded system, an external bus connects embedded processor cores, I/O peripherals, direct memory access (DMA) and off-chip memory. The power on the external bus makes up a significant portion of the overall power use in the system. In this paper, we will focus on the address and control bus power on the external bus. We have developed an external bus power model which monitors memory bus state transitions and models power-efficient bus arbitration schemes power. Our model allows us to consider performance/power trade-offs in managing off-chip memory accesses. We use an Analog Devices ADSP-BF533 multimedia system-on-a-chip embedded system as our target architecture to validate our model. By using more power-efficient external bus arbitration schemes, we find we can reduce overall power by as much as 18%.

Keywords: Power-Aware, External Memory, Bus Arbitration, Embedded System, Media Processor.

1 Introduction

There is a growing gap between the speed of microprocessors and the supporting off-chip memory systems. Also, the power associated with off-chip accesses can dominate the overall power budget. One approach to addressing both issues is to consider how best to schedule off-chip accesses. Due to the intrinsic capacitance of the bus lines, a considerable amount of power is required at the I/O pins of a system-on-a-chip processor when data has to be transmitted through the external bus [1,2]. The capacitance associated with the external bus is much higher than the internal node capacitance inside a microprocessor. For example, a low-power embedded microprocessor system like an Analog Devices ADSP-BF533 running at 500 MHz consumes about 374 mW on average during normal exection. Assuming a 3.65 V voltage supply and a bus frequency of 133 MHz, the average external power consumed is around 170 mW, which accounts for approximately 30% of the overall system power dissipation [3].

In modern CMOS circuit design, the power dissipation of the external bus is directly proportional to the capacitance of the bus and the number of transitions

B. Falsafi and T.N. Vijaykumar (Eds.): PACS 2004, LNCS 3471, pp. 95–106, 2005.

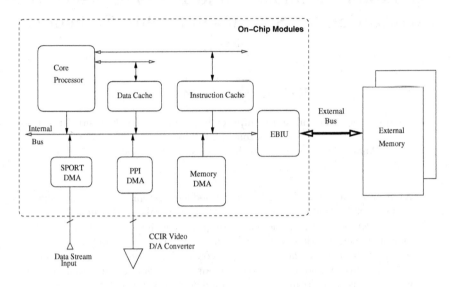

Fig. 1. Embedded Media System Architecture

(1 → 0 or 0 → 1) on bus lines [4,5]. In general, the external bus power can be expressed as:

$$P_{bus} = C_{bus} V_{ext}^2 f k \mu \tag{1}$$

In the above equation, C_{bus} denotes the capacitance of each line on the bus, V_{ext} is bus the supply voltage, f is the bus frequency, k is the number of bit toggles per transition on the full width of the bus, and μ is the bus utilization factor. This power equation is an activity-based model. It not only accounts for the dynamic power dissipated on the bus. It also includes the pin power that drives the signal I/O related to the external bus communication.

The techniques to minimize the power dissipation in buses are well explored in previous research [1,6,7]. The major approaches are utilizing the bus encoding to minimize the bus activity. Various mixed-bus encoding techniques (such as Gray code and redundant codes) were developed to save on bus power. Gray code addressing is based on the fact that address changes are often sequential and so using Gray codes to count switch the least number of signals on the bus. However, better performance can be obtained by using redundant code [1]. Many redundant codes have been proposed to add more signals on the bus line in order to reduce the number of transitions. Bus-invert coding [7] is one class of the redundant code. Bus-invert coding adds an INV signal on the bus to represent the polarity of the address on the bus. The INV signal value is chosen by considering how best to minimize the hamming distance between last address on the bus and the current one. Some codes can be applied to both the data and address bus, though some are more appropriate on to addresses. All power-saving codes are based on the assumption that the full width of the address bus are sent on each access. There are several issues related to external bus power that have not been properly addressed in previous work:

- An external bus contains three different components: control bus lines, address bus lines and bus data lines. The control bus consumes bus power, which was not considered in previous models.
- In some external memory modules (such as SDRAM), row and column address are shared on the address bus.
- Memory state transitions and stalls (such as page misses), which cause power and performance penalties, were not considered in the previous models.

The main contribution of this paper is the creation of an accurate external bus power estimation model, which overcomes the issues listed above. This new power model enables us to evaluate heuristics that can balance power/delay trade-offs associated with external bus data transfers. This paper is organized as follows. In section 2 we describe the target architecture for this work, which contains a system-on-a-chip processor, a 16 MB SDRAM and an external bus interface unit. Section 3 describes various bus arbitration algorithms for power and performance. In Section 4, we present our bus power modeling scheme, which considers both control signals and memory state transitions. Section 5 present power/performance results an MPEG-2 encoder and decoder. Finally, Section 6 presents conclusions.

2   Background

2.1   System Architecture

A typical system-on-a-chip embedded system includes many components: a high-speed processor core, hardware accelerators, a rich set of peripherals, direct memory access (DMA), on-chip cache and off-chip memory. The system architecture used in our study, which includes off-chip memory, is typical of current embedded platforms.

In modern multimedia applications, the requirements on processing throughput is increasing faster and faster. Today, for a D1 (720x480) video codec (encoder/decoder) media node, we need to be able to process 10 million pixels per second. This workload requires a RISC media processor for computation, devices with streaming-media capability via a parallel peripheral interface (PPI) for connecting high speed video and data converters, and synchronous serial ports (SPORT) for connecting to high speed telecom interfaces. The associated high data throughput requirements make it impossible to store all the data in on-chip memory or cache. Therefore, a typical multimedia embedded system usually has a high-speed system-on-a-chip microprocessor and a very large off-chip memory. The Analog Devices Blackfin family processors [8], the Texas Instrument OMAP [9] and the SigmaDesign EM8400 series [10] are all examples of low-power embedded media chip sets which share many similarities in system designs and bus structures. The system architecture in our study is extracted from those designs and is shown in Figure 1.

When trying to process streaming data in real-time, the greatest challenge is to provide enough memory bandwidth in order to sustain the necessary data rate. To insure sufficient bandwidth, hardware designers usually provide multiple buses in the system, each having different bus speeds and different protocols. An external bus is used to interface to the large off-chip memory system and other asynchronous memory-mapped devices. The external bus has a much longer physical length than other buses, and thus typically has much higher bus capacitance and greater power dissipation. The goal of work is to accurately model this power dissipation in a complete system power model so we can explore new power-efficient scheduling algorithms for the external memory bus.

2.2 External Bus Interface Unit

In the system design of Figure 1, there are four buses shown. Two buses that are clocked at the frequency of the processor are used to interconnect the processor and caches. There is one internal bus and one external bus, that are clocked at a slower frequency. The internal bus and external bus are bridged by an external bus interface unit (EBIU), which provides a glue-less interface to external memory.

There are two sub-modules inside the EBIU, a bus arbitrator and a memory controller. When the units (processor or DMA's) in the system need to have access to the external memory, they only need to make a request to the EBIU buffer, through the internal bus. The EBIU will read the request and handle the off-chip communication tasks through the external bus. Because of potential contention between users on the bus, arbitration for the external bus interface resources is required. The bus arbitrator grants requests based on a pre-defined order. Only one access request can be granted at a time. When a request has been granted, the memory controller will communicate with the off-chip memory directly based on the specific memory type and protocol. The EBIU can support SDRAM, SRAM, ROM, FIFOs, flash memory and ASIC/FPGA designs, while the internal units do not need to discriminate between different memory types. In this paper, we use multi-banked SDRAM as an example of memory technology and integrate SDRAM state transitions into our external bus model (our modeling framework allows us to consider different memory technologies, without changing the base system-on-a-chip model).

3   Bus Arbitration

The bus arbitration unit in the EBIU determines the sequencing of the load/store requests to SDRAM, with the goals of reducing contention and maximizing bus performance. The requests from each unit will be queued in the EBIU's wait queue buffer. When a request is not immediately granted, the request enters stall mode. Each request can be represented as a tuple (t, s, b, l), where t is the arrival time, s tells whether it is a load or store, b is the address of the block and l is the extent of the block. The arbitration algorithm schedules requests sitting in the wait queue buffer with a particular performance goal in mind. The algorithm needs to guarantee that bus starvation will not occur.

3.1 Traditional Algorithms

A number of different arbitration algorithms have been used in microprocessor system bus designs. The simplest algorithm is *First Come First Serve* (FCFS). In this algorithm, requests are granted the bus in the order of arrival. This algorithm simply removes contention on the external bus without any optimization and pre-knowledge of the system configuration. Because FCFS schedules the bus naively, the system performs poorly due to instruction and data cache stalls. No priority is given to cache accesses over DMA access (cache access tend to be more performance critical than DMA accesses). An alternative is to have a *Fixed Priority* scheme where cache accesses always have higher priority than DMA accesses. For different DMA accesses, peripheral DMA accesses will have higher priority than memory DMA accesses. This is needed because if a the peripheral device access is held off for a long period of time, it could cause the peripheral to lose data or get out of sync. The Fixed Priority scheme selects the request with highest priority in the waiting queue instead of the oldest. It may provide similar external bus performance as the FCFS algorithm, but the overall system performance should be better if the application is dominated by the cache accesses. For real-time embedded applications which are dominated by DMA accesses, cache accesses are tuned, and cache misses are infrequent. Cache fetches can be controlled to occur only at non-critical times using cache prefetching and locking mechanism. Therefore, for real-time embedded applications, FCFS and Fixed Priority scheme produce very similar external bus behaviors.

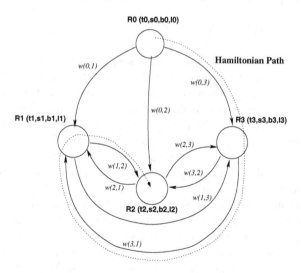

Fig. 2. Hamiltonian Path Graph

3.2 Power Aware Algorithms

To achieve an efficient external bus performance, FCFS and Fixed Priority are not sufficient. Power and speed are two major factors of bus performance. If a power-efficient arbitration algorithm is aware of the power and cycle costs associated with each bus request in the queue, each request can be scheduled so to better balance power/performance. The optimization target can be to minimize power P, minimize delay D or more generally to minimize $P^n D^m$. This problem can be formulated as a shortest Hamiltonian path (SHP) on a properly defined graph. The HamiltonianC path is a path in a directed graph that visits each vertex exactly once, without any cycles. The shortest Hamiltonian path is the Hamiltonian path that has the minimum weight. The problem is an NP-complete problem, and in practice, heuristic methods are used to solve the problem [11].

Let R_0 denotes the last request fulfilled on the external bus. $R_1, R_2, ... R_L$ are the requests in the wait queue. Each request R_i has four elements (t_i, s_i, b_i, l_i), representing the arrival time, operating type (load/store), starting address and access length. The bus power and delay are dependent on the current bus state and next bus state for each request. The current bus state is the state of the bus after the previous bus access has completed. $P(i, j)$ represents the bus power dissipated for request R_j, given R_i was the immediate past request. $D(i, j)$ is the bus delay between the time request R_j is submitted and the time request R_j is completed, where R_i was the immediate past request. The cost of selecting request R_j after request R_i can be formulated as $P^n(i, j)D^m(i, j)$. We can define a directed graph $G = (V, E)$ whose vertices are the requests in the wait queue, with vertex 0 representing the last request completed. The edges of the graph are all the pairs (i, j). The weight of each edge is weighted by $w(i, j)$, equal to the power delay product of processing request R_j after request R_i.

$$w(i, j) = P^n(i, j)D^m(i, j), n, m = 0, 1, ... \qquad (2)$$

The problem of optimal bus arbitration is equivalent to the problem of finding a Hamiltonian path starting from vertex 0 in graph G with a minimum traversal edge weight. Figure 2 describes a case when there are 3 requests in the wait queue. One of the Hamiltonian pathes is illustrated with a dot line. The weight for this path is $w(0, 3)+w(3, 1)+w(1, 2)$. For each iteration, a shortest Hamiltonian path will be computed to obtain the minimum weight path. The first request after request R_0 on that path will be the request selected in next bus cycle. After the next request is completed, a new graph will be constructed and a new minimum Hamiltonian path will be found.

Finding the shortest Hamiltonian path has been shown to be NP-complete. To solve the problem, we use heuristics. Whenever the path reaches vertex R_i, the next request R_k with minimum $w(i, k)$ will be chosen. This is a greedy algorithm, which selects the lowest weight for each step. The bus arbitration algorithm only selects the second vertex on that path. We avoid searching the full Hamiltonian path, and so the bus arbitration algorithm can simply select a request based on minimum $w(0, k)$ from request R_0. The complexity of that algorithm is $O(L)$. When $w(i, j) = P(i, j)$, arbitration can select to minimize power. When $w(i, j) = D(i, j)$, then we can minimize for delay. To consider the power efficiency, the power delay product can be used. Selecting different values for n and m change how we tradeoff power with delay using weights $w(i, j)$.

Table 1. SDRAM Commands Truth Table

Command	SMS	$SCAS$	$SRAS$	SWE	$SCKE$	$SA10$
PRECHARGE	low	high	low	low	high	high
ACTIVATE	low	high	low	high	high	
READ	low	low	high	high	high	low
WRITE	low	low	high	low	high	low
REFRESH	low	low	low	high	high	
NOP	low	high	high	high	high	

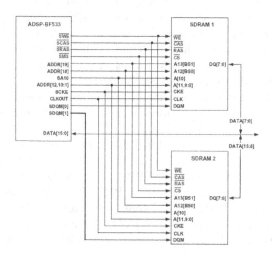

Fig. 3. SDRAM System Interface

To make sure the arbitration algorithm does not produce starvation, a time-out mechanism is added for the requests. The timeout values for cache and DMA are 100 and 550 cycles, respectively.

4 Power Modeling

To model power accurately, we are using the Analog Devices Blackfin frio-eas-rev0.1.7 toolkit. This model allows us to model power accurately, and has been validated with physical measurements as described in [12]. The power includes dynamic power to charge and discharge the capacitance along the external bus and the pin power to drive the bus current. The external bus power will be quite different if the memory technology is different. Today, there is almost no external bus power model that considers memory technology. In our model, we use SDRAM in our power model. The same approach is also applicable to other types of memory modules. The external bus power in each transaction will be the total number of pins that toggled on the bus. The power consumption includes the commands sent on the control bus, the row address and column address on the address bus and the data on data bus.

SDRAM is commonly used in cost-sensitive embedded applications that require large amounts of memory. The SDRAM model we are using is from Micron, MT48LC16M16A2 Synchronous DRAM. Figure 3 shows a block diagram of the SDRAM interface in the system architecture described in Section 2. The SDRAM can be organized as multiple banks. Inside each bank, there are many pages, which are selected by row address. The size of each page can be 1 KB, 4 KB or larger. The address inside one page is called the column address. The external bus is a multi-line bus, which can be grouped into three sub-buses: a control bus (including SCAS, SRAS, SWE, SCKE, SA10, DQM[1:0], BS[1:0]) which carries the SDRAM command signals and bank address, an address bus (including A[12:11], A[9:0]) which multiplexes the row address and the column address, and a data bus (DATA[15:0]) which transmits the loaded or stored data between the SDRAM interface and SDRAM.

The SDRAM operates on a command-by-command basis. Before every access to SDRAM, EBIU sends one or more commands on the control bus to signal to the SDRAM what the requested data is. The commonly used commands and their associated pin values are listed in Table 1. Between each pair of commands, a set amount of delay is required to meet the SDRAM specification. The delay cycles are preprogrammed into the SDRAM. t_{RAS} is the required delay between issuing an ACTIVATE command and a PRECHARGE command. t_{RP} is the required delay after a PRECARGE command. t_{RCD} is the delay between an ACTIVATE command and a READ/WRITE command. Column Address Strobe (CAS) latency t_{CAS} is the delay from a READ/WRITE command being issued to data ready.

There are various possibilities when accessing the SDRAM. Whenever a page miss occurs, the EBIU executes a PRECHARGE command followed by a bank ACTIVATE command, before executing the READ or WRITE command. This latency is called the *page miss penalty*. If there is a page hit, the READ or WRITE command can be transmitted immediately without requiring the PRECHARGE command. Figure 4 is a timing diagram for processing a read page miss operation. Some latency is required after the PRECHARGE and ACTIVATE commands. For SDRAM READ commands, there is a latency from the start of the READ command to the availability of data from the bus. This latency is always present for the first read in the burst and for any single read transfer. Subsequent read bursts do not have any latency, because those operations can be pipelined [8]. When a page miss occurs (about 70% of time for MPEG-2), more power and bus cycle delay are needed to complete the SDRAM access request. Another side effect occurs when the bus direction is switched (i.e., a READ after WRITE or WRITE after READ.) The bus controller pin needs time to turn around the bus, and is called the *bus turnaround penalty*.

In our bus model, we assume that the power to drive the control bus and address bus are the same. For each read/write request, we first determine the series of commands needed to complete that request. For each command, the

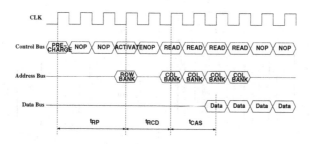

Fig. 4. Page Miss READ with PRECHARGE

bus state transitions, pins toggle and the utilization factor is recorded. Finally, the average bus power dissipation is calculated using Equation 1.

5   Experiments

5.1   Procedure

In our experimental study, we used a power model of the Analog Devices Black-fin family system-on-a-chip processors as our primary system model. We run code developed for ADSP-BF533 EZ-Kit Lite board using the VisualDSP++ toolset. This board provides a 500 MHz ADSP-BF533 microprocessor, 16 MB SDRAM, and CCIR-656 video I/O interface. Inside the ADSP-BF533 micropro-cessor, there are both L1 instruction and data caches. The instruction cache is 16 KB 4-way set associative. The data cache is 16 KB 2-way associative. Both caches use a 32 byte cache line size. The SDRAM module selected is Micron MT48LC16M16A2 16 MB SDRAM. The SDRAM interface connects to a 128 Mbit (x8) SDRAM devices to form one 16 MB of external memory. The SDRAM contains 4 banks, with a 1 KB page size. It also has following characteristics to match the on-chip SDRAM controller specification: supply voltage 3.3 V, oper-ating frequency 133MHz, burst length of 1, column address strobe (CAS) latency t_{CAS} 3 system clock cycles, t_{RP} and t_{RCD} 2 system clock cycles, refresh rate programmed at 4095 system clock cycles.

Our target workload is MPEG-2. We are using a real-time MPEG-2 en-coder and decoder source codes that include optimized Blackfin MPEG-2 codec libraries used in the ADI eMedia product. The input datasets used are the *cheer-leader* for encoding (the size is 720x480 and the format is interlaced video) and *tennis* for decoding (this image is encoded by the MPEG-2 reference encoder, the size is also 720x480, the format is progressive video). Both inputs are used heavily by the multimedia community.

In order to implement the external power model and be able to assess the impact of our power-efficient bus arbitration algorithm, we performed the follow-ing experiments. First, we modified the Blackfin instruction level simulator to include the system bus structure model and cache activity in order to generated a trace of external bus reference. In the second step, we built a separated EBIU

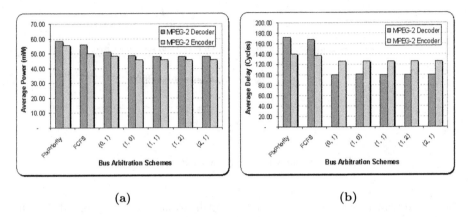

(a)   (b)

Fig. 5. MPEG-2 encoder and decoder external bus Power(a)/Delay(b) graph

module to simulate the external bus behavior capturing detailed SDRAM state transitions and a detailed bus arbitration module (as described in Section 3). The average bus power and performance are computed from the simulation results produced by our EBIU simulator.

5.2   Results

We consider the multimedia applications as representative workload for the system architecture in our experiment. The simulated applications were optimized Blackfin real-time MPEG-2 encoder and MPEG-2 decoder. Both applications includes many computational intensive functions and include both cache and DMA accesses to process D1 (720x480, 30fps) video content in less than 500 MIPS. For both applications, we ran the simulation for 170 million processor cycles to obtain the external bus traffic. There are 2.8 million EBIU requests recorded for the encoder and 2.9 million EBIU requests for the decoder.

There are seven different bus arbitration schemes in our simulation environment. We consider two traditional schemes: Fixed Priority, FCFS, and five power-efficient schemes. For Fixed Priority we assign the following priority order

Table 2. MPEG-2 Decoder Simulation Results

Arbitration Scheme	Utilization Factor (μ)	Average Power (P_{avg})	Average Delay (D_{avg})
FixedPriority	66.0%	58.47	171.94
FCFS	64.6%	56.05	168.51
(0, 1)	62.8%	50.89	100.66
(1, 0)	63.1%	48.57	101.52
(1, 1)	62.8%	48.14	100.42
(1, 2)	62.8%	48.18	100.63
(2, 1)	62.9%	48.23	100.78

Table 3. MPEG-2 Encoder Simulation Results

Arbitration Scheme	Utilization Factor (μ)	Average Power (P_{avg})	Average Delay (D_{avg})
FixedPriority	60.6%	55.85	140.36
FCFS	58.3%	50.19	137.93
(0, 1)	57.8%	48.63	125.69
(1, 0)	57.9%	45.91	126.18
(1, 1)	57.8%	45.94	125.87
(1, 2)	57.7%	45.99	125.83
(2, 1)	57.8%	45.95	125.98

from highest to lowest: instruction cache, data cache, PPI DMA, SPORT DMA, memory DMA. In the power-efficient schemes, we use the (n, m) format to represent them. n and m are the exponential numbers in Equation 2. Different n and m values will have different weights on power and delay. $(1, 0)$ is the minimum power scheme, $(0, 1)$ is the minimum delay scheme, and $(1, 1)$, $(1, 2)$, $(2, 1)$ have both power and delay in them. The MPEG-2 encoder and decoder simulation results are listed in Tables 2 and table3, respectively.

Figure 5 shows the power and delay of all seven arbitration schemes for both encoder and decoder. In the figure, we see that power-efficient schemes produce much lower power dissipation and and experience shorter bus delays than traditional schemes. Power-efficient scheme $(1, 1)$ enjoys an 18% power savings relative to a Fixed Priority scheme for both MPEG-2 encoder and decoder. For speed improvement, the power-efficient scheme $(1, 1)$ obtains a 40% reduction in cycles than the Fixed Priority scheme on MPEG-2 decoding, and a 10% reduction for MPEG-2 encoder. The speed improvement difference between the decoder and the encoder is due to the utilization factors. The encoder produces less traffic on the external bus than the decoder, therefore the average number of requests for the encoder in the wait queue is smaller than the number for the decoder.

Comparing the results across the power-efficient schemes, we can see that the performance differences are small, and that no one scheme provides significant advantages over the rest. The scheme $(1, 0)$, minimum power approach, is actually more favorable with regards to design implementation. It basically needs a Hamming distance (XOR) computation unit and a comparator. For each iteration, the arbitrator uses the Hamming distance computation unit to accumulate the power used for each request that is pending in the wait queue, and uses the comparator to select the minimum. For $0.13 \mu m$ CMOS technology and a 1.2 V power supply, an XOR transistor takes about 30 fJ to switch the transistor state in the slow N and slow P process corner. In our case, the number of transistors to implement the $(1, 0)$ arbitrator is on the order of 10^3. On average there are 2 requests in the wait queue and the request arrival interval is 60 cycles. The average power consumption is around 0.5 mW, which is much smaller than the power saving on the external bus.

6   Conclusions

With memory speed and bus capacitance continually increasing, accesses on the external bus consume more and more of the total power budget on a system-on-a-chip embedded system. This paper proposes a new external bus arbitration scheme that reduces bus power and delay. Our experiments are based on modeling a low-end embedded multimedia architecture while running real-time MPEG-2 encoding and decoding. Our results show that significant power and delay reductions can be achieved using power-efficient bus arbitration schemes.

References

1. Benini, L., De Micheli, G., Macii, E., Sciuto, D., Silvano, C.: Address bus encoding techniques for system-level power optimization. In: Proceedings of the conference on Design, automation and test in Europe, IEEE Computer Society (1998) 861–867
2. Panda, P.R., Dutt, N.D.: Reducing address bus transitions for low power memory mapping. In: Proceedings of the 1996 European conference on Design and Test, IEEE Computer Society (1996) 63
3. Analog Devices Inc. Norwood, MA: Engineer-to-Engineer Note EE-229: Estimating Power for ADSP-BF533 Blackfin Processors (Rev 1.0). (2004)
4. Givargis, T.D., Vahid, F., Henkel, J.: Fast cache and bus power estimation for parameterized system-on-a-chip design. In: Proceedings of the conference on Design, automation and test in Europe, ACM Press (2000) 333–339
5. P.P.Sotiriadis, Chandrakasan, A.: Low-power bus coding techniques considering inter-wire capacitances. In: Proceedings of IEEE Conference on Custom Integrated Circuits (CICC'00). (2000) 507–510
6. Lv, T., Henkel, J., Lekatsas, H., Wolf, W.: A dictionary-based en/decoding scheme for low-power data buses. IEEE Trans. Very Large Scale Integr. Syst. **11** (2003) 943–951
7. Stan, M., Burleson, W.: Bus-invert coding for low-power I/O. IEEE Transactions on Very Large Scale Integration (VLSI) Systems (1995) 49–58
8. Analog Devices Inc. Norwood, MA: ADSP-BF533 Blackfin Processor Hardware Reference (Rev 2.0). (2004)
9. Texas Instruments Inc. Dallas, Texas: OMAP5912 Multimedia Processor Device Overview and Architecture Reference Guide (Rev. A). (2004)
10. Sigma Designs, Inc. Milpitas, CA: EM8400: MPEG-2 Decoder for Set-top,DVD and Streaming Applications (Rev 01.09.03). (2003)
11. Rubin, F.: A search procedure for hamilton paths and circuits. J. ACM **21** (1974) 576–580
12. Vandersanden, S., Kaeli, D., Olivadoti, G., Gentile, R.: Developing energy-aware strategies for the blackfin processor. In: Proceedings of the Hight Performance Embedded Computing Conference. (September, 2004)

Context-Independent Codes for Off-Chip Interconnects

Kartik Mohanram and Scott Rixner

Rice University, Houston TX 77005, USA
{kmram, rixner}@rice.edu

Abstract. This paper introduces the concept of context-independent coding using frequency-based mapping schemes in order to reduce off-chip interconnect power consumption. State-of-the-art context-dependent, double-ended codes for processor-SDRAM off-chip interfaces require the transmitter and receiver (memory controller and SDRAM) to collaborate using current and previously transmitted values to encode and decode data. In contrast, the memory controller can use a context-independent code to encode data stored in SDRAM and subsequently decode that data when it is retrieved, allowing the use of commodity memories. In this paper, a single-ended, context-independent code is realized by assigning limited-weight codes using a frequency-based mapping technique. Experimental results show that such a code can reduce the power consumption of an uncoded off-chip interconnect by an average of 30% with less than a 0.1% degradation in performance.

1 Introduction

Modern embedded networking, video, and image processing systems are typically implemented as systems-on-a-chip (SoC) in order to reduce manufacturing costs and overall power and energy consumption. By integrating all of the peripheral functionality directly onto the same chip with the core microprocessor, both chip manufacturing and system integration costs can be lowered dramatically. In addition to cost, managing power and energy is a first order constraint that drives the design of embedded systems based on SoCs. However, most modern SoC-based embedded systems require more memory capacity than can reasonably be embedded into a single core. In such systems, the interconnect between the processor and external memory can consume as much or more power than the core itself. Even though external memory and its associated interconnect are major contributors to the overall power dissipation in SoC-based embedded systems, such systems will continue to require the memory capacity afforded by external memory into the foreseeable future. Therefore, it is essential to develop advanced memory controller architectures to reduce the power dissipation of external memories and the interconnect between the embedded SoC core and that memory.

Encoding data that is stored in memory can minimize the power consumed by the processor-memory interconnect. Dynamic power is consumed by the interconnect drivers when there are bit transitions. To minimize this power, double-ended, context-dependent codes such as the bus-invert code have previously been proposed. Double-ended codes encode data at the transmitter and decode it at the receiver. For a processor-memory interconnect, this implies that the SDRAM also needs to participate in such

B. Falsafi and T.N. Vijaykumar (Eds.): PACS 2004, LNCS 3471, pp. 107–119, 2005.
© Springer-Verlag Berlin Heidelberg 2005

codes. Context-dependent codes use the value last transmitted on the interconnect as well as the current data in order to encode the data to minimize transitions. For example, bus-invert coding either transmits the data value unchanged or its inverse, depending on which value minimizes transitions on the interconnect. If the SDRAM were modified to support such coding, bus-invert coding could reduce transitions on the interconnect by 22% on average.

In contrast, single-ended, context-independent codes are much simpler to implement, as they do not require modifications to the SDRAM. This paper introduces the concept of frequency-based, single-ended, context-independent codes for interconnect power reduction. The simplest frequency-based code simply remaps the input space based upon the measured or expected frequency of occurrence of each data value. Despite the fact that such a code is context-independent, and so does not account for possible switching on the interconnect, it is able to reduce the transitions on the interconnect by 28% on average. This simple code results in a larger power decrease on the interconnect than context-dependent bus-invert codes that are explicitly designed to minimize switching activity. Furthermore, frequency-based coding can also be used to augment limited-weight codes (LWCs). A limited-weight code maps the input data to a wider codeword in which the number of ones in the word is restricted. The proposed frequency-based assignment of codewords using a LWC can reduce transitions on the interconnect by an average of 30% over the uncoded case.

Frequency-based context-independent codes reduce interconnect power consumption without requiring the use of specialized SDRAM. Frequency-based codes are effective because frequently occurring values usually follow either themselves or other frequently occurring values on the interconnect. So, if the most frequently occurring values are all mapped to codewords that are close (Hamming distance-based) to each other, then switching activity can be minimized. In this manner frequency-based codes simply, but effectively, reduce dynamic power consumption on interconnects to commodity SDRAM.

The rest of this paper is organized as follows. The following section gives additional background on power dissipation within embedded systems, further motivating the need for new memory controller architectures. Section 3 introduces state-of-the-art coding techniques to reduce power consumption and discusses their limitations. In Section 4, the proposed frequency-based limited-weight codes for low power consumption are described. Section 5 describes a memory controller architecture for these coding techniques. Section 6 describes the experimental setup and the benchmarks used. Section 7 analyzes the performance of the proposed memory controller innovations on this set of embedded computing benchmarks. Section 8 concludes the paper.

2   Power Dissipation in Embedded Systems

The architecture of a modern SoC-based embedded system is presented in Figure 1. The SoC core has one or more simple processors, designed to provide enough computational capability for the application, integrated with some embedded memory and a variety of on-chip peripherals for data acquisition and connectivity. These systems also integrate SDRAM, since they frequently require more memory capacity—to buffer

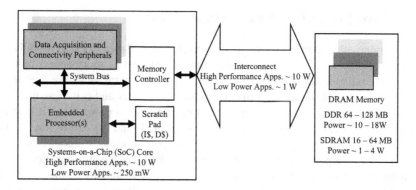

Fig. 1. Typical SoC Embedded System Architecture and Approximate Power Consumption

large data streams before either processing or forwarding them—than can reasonably be embedded into the SoC core.

Managing power dissipation and providing sufficient on-chip memory capacity are two major challenges in the design of such SoC-based embedded systems. The International Technology Roadmap for Semiconductors (ITRS) predicts that without significant architectural and design technology improvements, the power consumption of both high performance and low power SoC-based embedded systems will grow exponentially, easily exceeding power budgets [1,2]. Tethered embedded systems frequently have limited power budgets because of constraints on power delivery and cooling area available on peripheral buses. Mobile systems, in addition to requiring low power dissipation, are also constrained by battery life making energy consumption an important factor.

The annotations in Figure 1 show that, currently, the power dissipation of representative low power and high performance embedded systems is divided roughly equally among the SoC, the memory interconnect, and the external memory. Furthermore, while high performance embedded systems can dissipate an order of magnitude more power than low power systems, the relative power dissipation of the SoC core, the interconnect, and the memory remains similar. It is clear that in such systems, the external memory and interconnect can dissipate as much or more power than the SoC core. Thus, the memory system and the interconnect are candidates for techniques to reduce and manage power and energy.

Dynamic power is dissipated on a signal line of a bus whenever there is a transition on that line. A signal transition causes the drivers to actively change the value of the bus, which acts as a large capacitance. The drivers can also dissipate static power when they hold the bus at either logical 0 or logical 1, depending on the design of the drivers. It is possible to limit this leakage power for the low frequencies of operation commonly found in embedded systems by properly sizing the transistors within the bus drivers. Hence, static power dissipation is typically dwarfed by the dynamic power dissipation of the bus drivers in embedded systems. However, there are still situations in which static power dissipation cannot be ignored, including higher frequencies of operation and when there are voltage mismatches between the core and the memory.

3   Related Work: Coding to Reduce Power

The techniques used to reduce power dissipation in external memory systems fall roughly into three categories: low-power memory modes, external memory access reduction, and double-ended techniques. Most modern commodity memories have one or more low power modes of operation. It may be expensive to enter and exit these modes, but frequently the memory dissipates an order of magnitude less power when it is in these modes. Several techniques, such as those proposed in [3] and [4], can be used to determine when external memory should be powered down to minimize power dissipation without disrupting performance. Another way to reduce the power dissipation of external memories is to access them less frequently. These techniques use some combination of on-chip memory, caching, and code reorganization to allow the processing core to reduce the number of external memory accesses [5,6,7,8]. In turn, this reduces the power demands of the external memory when it is active and can also allow it to be put to sleep more frequently. The final set of techniques for reducing the power dissipation of external memories require cooperation between the memory controller and the memory. These techniques either encode data to minimize power dissipation across the interconnect or transmit additional information to the memory to enable it to access the memory array more efficiently [9,10,11,12,13].

The majority of data encoding schemes proposed in literature are not applicable to the off-chip interconnect between an SoC and external memory because they are double-ended, context-dependent codes. Double-ended codes require collaboration between the transmitter and receiver to transfer encoded data. In such state-of-the-art codes, the transmitter (i.e., the memory controller on the SoC) uses a potentially complex handshaking protocol to communicate with the receiver (i.e., a decoder in the memory), which has the ability to interpret these handshakes to decode the transmitted data. The roles of the coder and the decoder would be reversed when communicating in the opposite direction (i.e., a memory read). So, a potentially complex codec (coder-decoder) has to be present on both ends to successfully use these schemes. However, commodity SDRAMs do not have a built-in codec that is capable of communicating with the SoC core in this fashion.

Context-dependent coding schemes rely on inter-symbol correlation on successive data transfers to reduce power consumption. However, such schemes are not effective with commodity memory, as the memory cannot participate in the scheme. Therefore, any coding scheme using commodity memory must be able to unambiguously decode data read from the memory that was encoded when it was written to the memory. If inter-symbol correlation information is used when writing the data, then that information is not available upon reading the data, since there is no guarantee that data will be read in exactly the same order it is written. Some context-dependent coding schemes, such as those that use an XOR decorrelator, do not include enough information in the codeword to unambiguously recover the original data without the context information. However, other context-dependent schemes, such as bus-invert coding, produce codewords that can be decoded without context information. Even then, such schemes will only minimize power when writing to the memory, as the data will be read in a different context than it was written. Therefore, context-dependent codes are almost exclusively used in situations where both the transmitter and receiver can participate in the code.

That way, the context information can be used to decode the transferred data before it is stored. If the data is retrieved later, it is re-encoded and transferred based on the context at that time.

The most popular and easy-to-implement double-ended context-dependent code reported in literature in the bus-invert code [14]. The bus-invert code is a context-dependent, double-ended code since it computes the Hamming distance between the currently encoded data value on the bus and the next data value. If the Hamming distance exceeds $\lceil \frac{n}{2} \rceil$, then the transmitter inverts the next value transmitted on the bus. An additional line on the bus indicates whether the data is inverted or not, allowing the receiver to unambiguously decode the transmitted data. In this manner, an n-bit value can be transmitted over an $n + 1$-bit bus with at most $\lfloor \frac{n+1}{2} \rfloor$ transitions. Without such coding, an n-bit value could cause as many as n transitions over an n-bit bus. For example, if the current value on the bus is 0000, and the next value to be transfered is 0001, then the Hamming distance between the values is 1. Therefore, 0001 will be transmitted over the bus with the invert bit set to 0, indicating the data is not inverted. However, if the current value on the bus is 1111 instead, the Hamming distance between the values is 3 and hence 1110 is transmitted with the invert bit set to 1. In this manner, each information symbol in the n-bit input space maps to two codewords. The codeword that minimizes switching activity on the interconnect is chosen for transmission to reduce power consumption. Bus-invert is thus not a one-to-one mapping, i.e., it is not a context-independent code. The bus-invert codewords for all the information symbols on a 4-bit wide data bus are shown in column 2 of Table 1.

Many other context-dependent, double-ended codes have been proposed. One such code is based on the use of a decorrelator, which XOR's the data to be transmitted with the previous value transmitted across the bus [15,16]. The receiver must then recover the actual value by undoing the XOR operation. Further reductions can be achieved by exploiting information about the frequency of occurrence of particular data values on the bus. In [17], a decorrelator was combined with a one-hot encoding of the 32 most frequently occurring values. Like bus-invert, such frequent value encoding is still a context-dependent, double-ended code because of the use of the decorrelator. The transmitter first decides if the data value is one of the most 32 frequently occurring values. If so, it is one-hot encoded. A one-hot code on a n-bit wide bus is a coding scheme where exactly one out of n bits is set to one. At the word-level, 32 codewords are available and hence 32 frequently occurring values can be encoded leaving the remaining values unencoded. ** Note that an additional bit is needed to indicate whether or not the data is one-hot encoded to the receiver. The result of one-hot encoding is then passed through the decorrelator prior to transmission across the bus. The receiver must recover the actual value by undoing the XOR and one-hot encoding transformations. The final column of Table 1 shows the one-hot codeword assignments, based on the frequency distribution in column 4, for frequent value coding on a 4-bit wide data bus. Note that only 4 most frequently occurring values (1101, 1001, 0111, 0100) are one-hot encoded. In practice, these values would also be XOR'ed with the previous value

---

** The reported code also ignored values 1-16 and performed equality tests before transmission, the details of which are excluded for brevity [17]. Nevertheless, our experimental setup implemented the best scheme reported in [17] that includes some of these features.

Table 1. Comparison of Different Codes

Information Symbols	Bus-invert Coding [14]	2-LWC Code [13]	Frequency (%)	Freq.-based remapping	Freq.-based 2-LWC	Freq. value coding [17]
0000	0 0000 / 1 1111	0 0000	6.7	1001	0 0110	0 0000
0001	0 0001 / 1 1110	0 0001	5.6	0101	0 1010	0 0001
0010	0 0010 / 1 1101	0 0010	4.7	1011	1 1000	0 0010
0011	0 0011 / 1 1100	0 0011	6.9	1010	0 0011	0 0011
0100	0 0100 / 1 1011	0 0100	7.6	0100	0 0100	1 1000
0101	0 0101 / 1 1010	0 0101	7.0	1100	1 0000	0 0101
0110	0 0110 / 1 1001	0 0110	4.0	1101	1 0010	0 0110
0111	0 0111 / 1 1000	1 1000	8.1	0010	0 0010	1 0100
1000	0 1000 / 1 0111	0 1000	4.8	1110	0 1001	0 1000
1001	0 1001 / 1 0110	0 1001	8.4	0001	0 0001	1 0010
1010	0 1010 / 1 0101	0 1010	5.9	0011	0 1100	0 1010
1011	0 1011 / 1 0100	1 0100	4.0	0111	1 0100	0 1011
1100	0 1100 / 1 0011	0 1100	6.6	0110	0 0101	0 1100
1101	0 1101 / 1 0010	1 0010	8.5	0000	0 0000	1 0001
1110	0 1110 / 1 0001	1 0001	3.7	1111	1 0001	0 1110
1111	0 1111 / 1 0000	1 0000	7.5	1000	0 1000	0 1111

transmitted across the bus by the decorrelator. Like bus-invert, frequent value encoding does not use a one-to-one mapping, as a particular data value can map to many encoded values depending on the previous data transmitted across the bus.

4   Context-Independent, Single-Ended Codes

This section introduces and develops a class of context-independent, single-ended coding schemes for embedded applications. These coding schemes are split into two phases.

During the first phase, a set of codewords is generated. In the second phase, each information symbol is assigned to a unique codeword. These assignments are determined purely by frequency-based metrics without using any local context information. Such codes have several advantages. First, they significantly lower power consumption on the interconnect between the SoC and the memory modules. Second, they are single-ended, i.e., they do not require the SDRAM to participate in the coding-decoding process. A codec is only required in the memory controller on the SoC. Last, they have negligible impact on performance during the coding-decoding process.

4.1   Limited-Weight Codes (LWCs)

One class of context independent codes that meet all the above requirements are limited-weight codes (LWCs) [13]. Consider a k-bit wide data bus with 2^k information symbols. A m-LWC is a one-to-one mapping where every word in the 2^k input space maps to a codeword such that the Hamming weight (i.e., the number of ones in the codeword) is less than or equal to m. Since the source entropy must remain unchanged, i.e., since every information symbol must have a unique codeword, the following inequality must be satisfied by all m-LWCs:

$$\binom{n}{0} + \binom{n}{1} + \binom{n}{2} + \cdots + \binom{n}{m} \geq 2^k \tag{1}$$

Here, n is the minimum number of bits ($m \leq n$) such that the inequality is satisfied, i.e, n determines the width of the bus needed to implement a m-LWC. Note that the inequality is only satisfied when $n \geq k$. A perfect m-LWC satisfies Equation 1 above with equality, i.e., all the codewords of length n with weight less than or equal to m are used in the mapping. For example, a 4-LWC where k equals 8 is a perfect 4-LWC when the codeword bus width n equals 9. The information symbols and the corresponding codewords for the perfect 2-LWC when k equals 4, obtained using the generation technique presented in [13], are presented in columns 1 and 3 respectively in Table 1.

4.2   Frequency-Based Codes

Frequency-based codes are a class of codes where a context-independent mapping between information symbols and codewords is achieved by assigning information symbols that have the highest probability of occurrence to codewords with minimum weight. As discussed in Section 3, one way to use frequency would be to one-hot encode a small set of frequently occurring values at the word-level to achieve power savings. However, such an approach does not use the codeword space efficiently since only 32 out of 2^{32} values can actually be encoded by this scheme.

A more efficient way to use frequency is to remap information symbols to be transmitted on the n-bit wide bus to codewords on a n-bit wide bus, i.e., through a permutation. Such a remapping can be statically determined by an analysis of several traces in the application space. The frequency distribution of information symbols from each application could be used for that application, or the frequency distributions for a set of applications could be combined to produce a global frequency distribution. The information symbols are then ranked in descending order of frequency of occurrence,

and remapped to codewords in increasing order of weight. Thus, the information symbol that occurs most frequently on the bus is remapped to the codeword with the least
weight. In practice, such a remapping would have to occur at the byte-level since word-
level remapping is impractical.

The frequency distribution given in Table 1 can be used to perform such a remap
ping of the information symbols in the table. The information symbols and the corresponding codewords for a frequency-based remapping of the 4-bit information space are
presented in columns 1 and 5 respectively in Table 1. The frequency distribution used
to generate this remapping is presented in column 4 in the same table. For example,
1101 which is the most frequently occurring information symbol would be remapped
to 0000.

4.3   Frequency-Based m-LWCs

The biggest handicap of m-LWCs is that the one-to-one mapping is statically determined without any knowledge of the characteristics of the input space. The main contribution of this paper is to combine the advantages of frequency-based codes with LWCs
to produce a context-independent single-ended mapping. By combining m-LWCs with
frequency-based coding, the distribution of information symbols is analyzed to produce
a context-independent mapping that, while statically determined, exploits an a priori
knowledge of the distribution of information symbols. Frequency-based m-LWCs thus
leverage the advantages of both frequency-based coding and limited-weight coding. It
is a departure from conventional types of codes that seek to explicitly minimize transitions using the state of the bus. The generation of the codewords is separated from
the mapping process, and the best of both techniques is harnessed to realize practical
context-independent single-ended codes.

The frequency-based mapping encodes information symbols with the highest frequency of occurrence to LWC codewords with the least weight. A simple frequency-
based mapping from a 2-LWC to a 4-bit information space, using the frequency distribution in column 4 of Table 1 is presented in column 6 of the same table.

5   Memory Controller Architecture

Figure 2 shows the architecture of a memory controller for embedded systems. As the
figure shows, memory requests arrive on the system bus. At this point, if the memory
request is a write, the data will be encoded by the context-independent encoder before
it is placed in a queue within the memory controller. The SDRAM controller within
the memory controller then issues the appropriate commands to the SDRAM in order
to satisfy each pending request in the queue. Finally, if the memory operation is a read
from the SDRAM, the data can be decoded by the context-independent decoder before
being returned to the core over the system bus.

A context-independent codec does not need to be near the pins. Rather, the data
can be encoded and decoded anywhere within the memory controller because only the
actual data being encoded or decoded is needed to perform the encoding or decoding.
This makes it convenient to encode write data before it is placed in the memory queue,

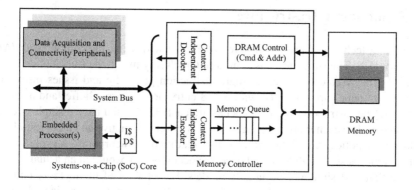

Fig. 2. Memory controller architecture for embedded systems

thereby minimizing any latency penalties. It is entirely possible that the latency of encoding write data can be hidden by long latency SDRAM operations that must occur before the data can cross the pins anyway. Similarly, read data can be decoded as it is sent to the system bus. Again, the decoding latency could possibly be hidden by arbitration delays for the system bus. For the 30 benchmark programs from the MiBench suite that will be explored in Section 7, an extra cycle latency penalty for decoding results in less than a 0.1% performance penalty on average.

A context-independent codec can be implemented in multiple ways. In the most general case, such as frequency-based coding, a lookup-table is the most efficient mechanism. To encode or decode bytes, a 256-entry table would be required with either 8 or 9 bit entries, depending on the code. For performance, it is likely that multiple identical tables would be required, one for each byte that can be transferred on the system bus in a given cycle. If the code is fixed, then the lookup-tables can be compact ROM tables. To provide the flexibility to change the code, however, it is likely that the lookup-tables would have to be SRAM structures. Combinational logic can be used to implement more regular context-independent codes. For example, the limited-weight code described in [13] could be implemented using a simple population count and possible inversion.

Finally, many of the codes discussed here increase the size of the data by adding an additional bit for every byte. This would increase the datapath width of the memory controller, the width of the processor-memory interconnect, and the width of the SDRAM. Obviously, this additional bit can increase power consumption, but the objective of these codes is to reduce power consumption by limiting the number of transitions, so usually this is not an issue in the memory controller or the processor-memory interconnect as will be shown in Section 7 (all results include the transitions on this additional wire, as appropriate). However, widening the SDRAM is potentially problematic. Many SRAMs designed for embedded systems have 9-bit bytes. And Samsung is starting to introduce SDRAMs of that nature, as well [18]. The wider Samsung SDRAMs consume 6–8% more current than their normal counterparts. However, this is assuming a regular data pattern. In practice, the reduction in switching activity achieved by these codes can more than offset this increase.

6   Simulation Infrastructure

The coding techniques presented here were evaluated using the SimpleScalar/ARM simulator [19]. The simulator was configured to closely match the Intel Xscale processor [20]. The Xscale can fetch a single instruction per cycle and issues instructions in order. Branches are predicted with a 128 entry branch target buffer and a bimodal predictor. The instruction and data caches are each 32 KB and have a single cycle access latency. The caches are configured as 32-way set associative and use a round-robin replacement policy. SimpleScalar was also modified to incorporate a cycle accurate SDRAM model so that all SDRAM accesses occur as they would in an actual system.

The SDRAM simulator accurately models the behavior of the memory controller and the SDRAM. The SDRAM model simulates all timing parameters, including command latencies, all required delays between particular commands, and refresh intervals. The memory controller within the simulator obeys all of these timing constraints when selecting commands to send to the SDRAM, thereby accurately representing the sequence of data transferred over the processor-memory interconnect. The simulator is configured to model a 75 MHz, 512 Mb Micron MT48LC32M16A2-75 single data rate SDRAM [21].

The bit transitions on the interconnect for the encoded and unencoded data transfers was calculated as the SDRAM is accessed. This faithfully models the bit transitions that would occur on the data bus in the appropriate order.

The MiBench embedded benchmark suite was used to evaluate the proposed coding techniques [22]. Thirty applications are used from the suite with their large input sets. While still small, the large inputs are more representative of actual workloads. The applications span the automotive, consumer, networking, office, security, and telecomm domains.

7   Results

Table 2 shows the average reduction in transitions on the processor-memory interconnect for nine coding strategies when compared with the baseline uncoded case. The

Table 2. Average reduction in transitions

Code		Reduction (%)
Context-dependent Double-ended	Bus Invert	21.8
	Self FV32 with Decorrelator	38.7
Context-independent Single-ended	Self FV32	17.8
	Self FV8	15.5
	4-LWC	13.9
	Self 8-LWC	28.2
	Global 8-LWC	22.4
	Self 4-LWC	30.3
	Global 4-LWC	25.1

first two codes in the table are context-dependent, double-ended codes. As described in Section 3, bus invert coding is the simplest and most popular such code, and FV32 with a decorrelator one-hot encodes the 32 most frequently occurring values (for each benchmark) and uses a decorrelator to significantly reduce switching activity. As the table shows, both context-dependent, double-ended codes perform quite well, reducing transitions on the interconnect by 21.8% and 38.7%, respectively.

The remaining seven codes are all context-independent, single-ended codes that can be implemented entirely within the memory controller without specialized SDRAM. FV32 and FV8 simply one-hot encode the 32 most frequently occurring word values or the eight most frequently occurring byte values to form a code as presented in Table 1. These codes are labeled "self", as each benchmark uses the most frequently occurring values from that benchmark. As the table shows, these codes perform poorly, yielding only a 17.8% and 15.5% reduction in switching activity. Therefore, such a one-hot encoding strategy relies heavily on a context-dependent, double-ended decorrelator to reduce transitions on the interconnect.

4-LWC is the original limited-weight code, presented in Section 4, which uses nine bits per byte to encode all byte values with at most four bits set. This code reduces switching activity by 13.9% on average.

The final four limited-weight codes use the frequency-based assignment scheme presented in this paper. "Self" and "Global" refer to whether each benchmark's own frequency distributions were used to assign codewords for that benchmark or all benchmarks used the same codewords derived from the frequency distributions of all benchmarks. The 8-LWC codes use eight bits to encode each byte, with up to eight bits set, yielding a simple remapping. The 4-LWC codes again encode each byte using nine bits with most four bits set. As the table shows, these codes are able to reduce transitions on the interconnect by 22.4–30.3% on average. As would be expected, the codes which use the frequency distributions for each benchmark individually yield higher reductions, by about 5–6%. These results show that when using limited-weight codes, the assignment strategy is critical. Furthermore, the penalty of using an extra wire for the 4-LWC codes is more than offset by the effectiveness of such codes (the results include the switching on the additional wire).

8   Conclusions

State-of-the-art coding techniques to reduce power consumption across the processor-memory interconnect have traditionally used context-dependent, double-ended techniques. This requires specialized memory that can participate in such codes. This paper introduced viable context-independent, single-ended codes that are competitive with these state-of-the-art codes, but can be used with commodity memory.

The proposed codes are effective at reducing power without degrading performance for thirty applications from the MiBench embedded benchmark suite with their large input sets. Frequency-based remapping codes require no augmentation to the memory bus or modules. These codes reduce the transitions of the uncoded data stream by 28.2% on average, and minimally impact performance (by less than 0.1% on average) across the set of thirty benchmark programs. Frequency-based 4-LWCs reduce the transitions of the uncoded data stream by 30.3% on average, improve on context-dependent double-ended bus-invert codes by 10.9% on average, and minimally impact performance (by less than 0.1% on average) across the set of thirty benchmark programs.

This paper has shown that both the type of code used and the assignment scheme for that code are important. Limited-weight codes by themselves are ineffective. Similarly, using frequency information without limited-weight codes yields an inefficient code that is also ineffective. However, using frequency information to assign limited-weight codes minimizes transitions to a greater extent than any other context-independent, single-ended code. Furthermore, such codes sometimes outperform context-dependent, double-ended codes that cannot be used with commodity SDRAMs. Since embedded systems continue to use commodity memories and the processor-memory interconnect is a dominant consumer of power in such systems, the coding techniques presented here can significantly improve the overall power efficiency of modern embedded systems.

References

1. Edenfeld, D., Kahng, A.B., Rodgers, M., Zorian, Y.: 2003 technology roadmap for semiconductors. *IEEE Computer* **37:1** (2004)
2. International technology roadmap for semiconductors (2003)
3. Delaluz, V., Kandemir, M., Vijaykrishnan, N., Sivasubramaniam, A., Irwin, M.: DRAM energy management using software and hardware directed power mode control. In: *Proceedings of the International Symposium on High-Performance Computer Architecture* (2001)
4. Fan, X., Ellis, C., Lebeck, A.: Memory controller policies for DRAM power management. In: *Proceedings of the International Symposium on Low Power Electronics and Design* (2001)
5. Catthoor, F., Wuytack, S., DeGreef, E., Balasa, F., Nachtergaele, L., Vandecappelle, A.: *Custom Memory Management Methodology: Exploration of Memory Organisation for Embedded Multimedia System Design*. Kluwer Academic Publishers (1998)
6. Kulkarni, C., Catthoor, F., DeMan, H.: Code transformations for low power caching in embedded multimedia processors. In: *Proceedings of the International Parallel Processing Symposium* (1998)
7. Kulkarni, C., Miranda, M., Ghez, C., Catthoor, F., Man, H.D.: Cache conscious data layout organization for embedded multimedia applications. In: *Proceedings of the Design Automation and Test in Europe Conference* (2001)
8. Panda, P.R., Dutt, N.D., Nicolau, A.: On-chip vs. off-chip memory: the data partitioning problem in embedded processor-based systems. *ACM Transactions on Design Automation of Electronic Systems* **5:3** (2000)
9. Benini, L., Macii, A., Poncino, M., Scarsi, R.: Architectures and synthesis algorithms for power-efficient bus interfaces. *IEEE Transactions on Computer-aided Design of Integrated Circuits and Systems* **19:9** (2000)
10. Ramprasad, S., Shanbag, N.R., Hajj, I.N.: A coding framework for low power address and data buses. *IEEE Transactions on VLSI Systems* **7:2** (1999)
11. Sotiriadis, P., Chandrakasan, A.: A bus energy model for deep sub-micron technology. *IEEE Transactions on VLSI Systems* **10:3** (2002)
12. Sotiriadis, P., Tarokh, V., Chandrakasan, A.P.: Energy reduction in VLSI computation modules: An information-theoretic approach. *IEEE Transactions on Information Theory* **49:4** (2003)
13. Stan, M.R., Burleson, W.P.: Low-power encodings for global communication in CMOS VLSI. *IEEE Transactions on VLSI Systems* **5:4** (1997)

14. Stan, M.R., Burleson, W.P.: Bus invert coding for low power I/O. *IEEE Transactions on VLSI Systems* **3:1** (1995)
15. Benini, L., DeMicheli, G., Macii, E., Sciuto, D., Silvano, C.: Asymptotic zero-transition activity encoding for address busses in low-power microprocessor-based systems. In: *Proceedings of the Great Lakes Symposium on VLSI* (1997)
16. Musoll, E., Lang, T., Cortadella, J.: Exploiting locality of memory references to reduce the address bus energy. In: *Proceedings of the International Symposium on Low Power Electronics Design* (1997)
17. Yang, J., Gupta, R., Zhang, C.: Frequent value encoding for low power data buses. *ACM Transactions on Design Automation of Electronic Systems* **9:3** (2004)
18. Samsung: 256/288 Mbit RDRAM K4R571669D/K4R881869D data sheet, version 1.4 (2002)
19. Austin, T., Larson, E., Ernst, D.: SimpleScalar: An infrastructure for computer system modeling. *IEEE Computer* **35:2** (2002)
20. Clark, L.T., Hoffman, E.J., Miller, J., Biyani, M., Liao, Y., Strazdus, S., Morrow, M., Velarde, K.E., Yarch, M.A.: An embedded 32-b microprocessor core for low-power and high-performance applications. *IEEE Journal of Solid-state Circuits* **36:11** (2001)
21. Micron: 512Mb: x4, x8, x16 SDRAM MT48LC32M16A2 data sheet (2004)
22. Guthaus, M.R., Ringenberg, J.S., Ernst, D., Austin, T.M., Mudge, T., Brown, R.B.: MiBench: A free, commercially representative embedded benchmark suite. In: *IEEE Annual Workshop on Workload Characterization* (2001)

Dynamic Processor Throttling for Power Efficient Computations

Masaaki Kondo and Hiroshi Nakamura

Research Center for Advanced Science and Technology, The University of Tokyo,
4-6-1 Komaba, Meguro-ku, Tokyo, Japan
{kondo, nakamura}@hal.rcast.u-tokyo.ac.jp

Abstract. We propose a novel hardware-based DVS technique called dynamic processor throttling (DPT) for power efficient computations. DPT focuses on the performance balance between the processor and main memory. When a performance imbalance is detected, DPT tries to redress the imbalance by setting the clock frequency and supply voltage of the processor to a well-balanced point.

This paper describes the micro-architecture mechanisms of DPT and shows the evaluation results on energy saving and performance compared with a conventional cache-miss-driven DVS technique. The results reveal that DPT can reduce 17% of the energy with a 3.4% performance degradation and DPT surpasses the conventional technique in both performance and energy.

1 Introduction

Reducing power/energy consumption has become a crucial issue for not only battery powered and embedded processors but also high performance microprocessors because power and thermal problems are certainly a key factor in limiting processor performance. To satisfy the power/thermal constraint of a chip, low-power architectural techniques are indispensable especially for future microprocessors.

Dynamic Voltage Scaling (DVS) has become attractive for high-performance and low-power computing. DVS selectively scales down the supply voltage during computations if the processing demand is not heavy. Because dynamic power consumption in a CMOS circuit scales quadratically with the supply voltage, a significant amount of power is saved by lowering the supply voltage. However, lowering the voltage degrades the speed of the circuit, and thereby the clock frequency must be lowered, which has a negative impact on performance.

Much DVS-related research has been carried out for power saving. In real-time systems especially, many techniques have been well studied [1, 2, 3, 4] to exploit the excessive computation power and to scale down the supply voltage under the constraint of a time deadline. Another approach is scaling down the supply voltage when the processor is likely to stall due to a cache miss. Because the performance gap between the processor and main memory is very large, processors waste significant amounts of time while waiting for data from the main memory. The performance penalty due to low clock frequency can be masked by the stall time. Therefore, this approach has a significant opportunity for power/energy saving without performance degradation.

B. Falsafi and T.N. Vijaykumar (Eds.): PACS 2004, LNCS 3471, pp. 120–134, 2005.

In such a DVS method, the application behavior is analyzed and the voltage is controlled either by the compiler [6, 11] or the hardware (micro-architecture) [7, 8, 9]. In compiler-based approaches, compilers statically identify the program regions where the performance does not degrade, even if the clock frequency is lowered. However, the static analysis does not predict perfectly the behavior of the programs. For example, cache behavior is difficult to analyze because it is affected by the data set of programs. On the contrary, hardware-based approaches can capture the dynamic behavior of program execution. Therefore, a micro-architectural approach is more attractive for a memory-bound-based DVS method.

In the hardware-based approaches, the cache-miss-driven methods proposed in [7, 8] have been well studied. They lower the supply voltage when L2 cache misses occur and switch the voltage back to high when one of the outstanding misses returns. In this paper, we propose a different hardware-based DVS technique called *Dynamic Processor Throttling (DPT)*, which detects the performance imbalance between the processor and main memory and redresses the balance. DPT tries to balance the throughput of a processor with that of the main memory by setting the clock frequency and supply voltage of the processor. When the memory performance dominates the execution time, the throughput of the processor is balanced by lowering the processor's frequency. Because the throughput balance dynamically changes within a program, its execution is divided into several time intervals and the frequency and voltage are reset at every interval. DPT is superior to conventional cache-miss-driven methods in the following aspects.

- Energy consumption is further reduced by setting the clock frequency and supply voltage at the balanced point.
- Because voltage switching is less frequent than cache-miss-driven methods, the performance and energy overheads are suppressed.

Due to the above points, DPT has the potential to save more energy with less performance penalty than conventional methods.

In DPT, the most important issue is how to detect the performance imbalance between the processor and main memory. Instructions Per Cycle (IPC) is one way because IPC is the simplest indicator of processor activity. In our studies, however, IPC does not always reflect the imbalance. Therefore, we propose another method for detecting the performance imbalance.

This paper is organized as follows. The next section describes the related work. In Section 3, we propose our DPT method and compare it with a miss-driven DVS technique. Section 4 describes the evaluation methodology and assumptions. The results are presented in Section 5. Finally, we conclude in Section 6.

2   Related Work

The DVS techniques have been well studied and adopted in several commercial processors so far.

Marculescu [7] proposed the cache-miss-driven DVS technique in which the supply voltage is lowered when the processor detects L2 cache misses. Li et al. [8] proposed *Variable Supply-Voltage scaling (VSV)*, which is an extension of the cache-miss-driven method. In this work, the performance penalty and energy overhead of the frequency/voltage transition are taken into consideration. Moreover, the voltage scaling is controlled with monitoring instruction level parallelism (ILP). VSV does not scale down the supply voltage when ILP is high because significant performance loss would ensue from lowering the voltage for such a program region.

These methods reduce power/energy consumption by scaling down the supply voltage when the processor stalls and waits for data from the main memory. Although the basic concepts of these methods are similar to our DPT, DPT focuses on the imbalance of the processor and main memory performance and tries to balance them with each other. Thus, this paper shows that DPT is a more energy-efficient solution with less performance penalty.

Stanley-Marbell et al. [9] proposed the Power Adaptation Unit (PAU), which is used to control the operating voltage of various system components such as CPU. The PAU scales the supply voltage to reduce power consumption without incurring more than a prescribed performance penalty. PAU attempts to identify dynamic program regions for voltage scaling, including regions with an imbalance in memory and CPU activity. However, PAU relies on a table which keeps track of the program execution status to determine the program regions. It may incur large energy overhead. Moreover, the case of an out-of-order superscalar processor is not evaluated. On the other hand, our DPT needs a very simple hardware structure only and our target is out-of-order superscalar processors.

Besides micro-architectural DVS techniques, there have been proposed several methods that optimize the supply voltage for a given program in the compilation phase by static compiler analysis or profiling [6, 11]. These methods statically identify the program regions where the performance does not degrade even if the clock frequency is lowered. However, it is difficult to completely analyze the behavior of the program execution, especially cache performance, statically because it is affected by the data set of the program. On the contrary, hardware-based approaches can capture the dynamic behavior of the program execution.

The DVS method can be applied more finely for Multi Clock Domain (MCD) micro-architecture [12, 13, 14]. It divides an entire processor chip into individual clock domains using the globally asynchronous locally synchronous (GALS) technique. There is the possibility of further reduction of power consumption by this fine-grain optimization. This work, however, does not consider the performance imbalance between the processor chip and main memory as much. Our proposed method is orthogonally applied to their MCD architecture.

Sasanka et al. [4] proposed an energy-driven architecture adaptation method for multimedia applications. This method adapts the voltage setting and architectural structure, including the instruction window size, to the time deadline and other requirements of applications. However, it relies on the characteristic of multimedia applications, that is, frame-based computation. On the contrary, our target is more general applications.

Bahar and Manne [5] proposed Pipeline Balancing (PLB), which dynamically tunes the resources of a processor to the needs of a program. PLB monitors IPC variations to determine when to enter or exit the low-power mode. In the low-power mode, the issue rate is reduced to disable some processor resources. In their work, however, DVS is not considered.

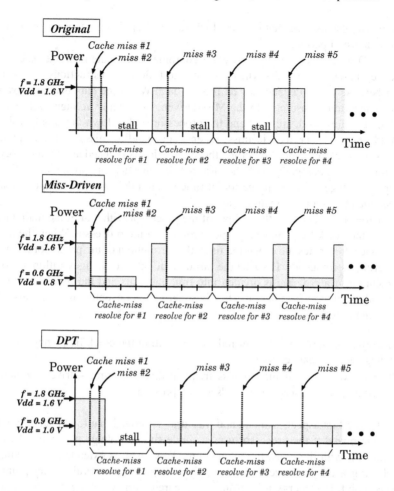

Fig. 1. Power consumption for three processors

3   Dynamic Processor Throttling: DPT

3.1   Overview of DPT

We propose a micro-architectural method called Dynamic Processor Throttling (DPT) which detects a performance imbalance between the processor and main memory, and adjusts the processor's supply voltage and clock frequency to redress the balance. Unlike the cache-miss-driven DVS method, which alternately uses two supply voltages, DPT chooses a supply voltage from several setting points.

Fig. 1 illustrates an example of the power consumption of three processors: an original processor without DVS (*Original*), a cache-miss-driven DVS processor (*Miss-Driven*), and our proposed DPT processor (*DPT*). In this figure, we assume the instructions are issued out-of-order, the cache is non-blocking, and one cache-miss request (that is, the data transfer between the processor and main memory) can be served at a moment. We also assume that perfect clock gating is applied so that no power is consumed in the stall period.

In the Original scenario, the processor continues to execute instructions even after the cache misses, but it stalls when none of the following instructions can be issued further because of the dependency on missed data. When the data becomes available, then its execution is resumed. In the Miss-Driven, two supply voltages, referred to as high-Vdd and low-Vdd, are used and the processor scales down the supply voltage to low-Vdd when it detects a cache miss. When an outstanding miss returns, the processor switches back to high-Vdd. Therefore, the stall period is reduced and total energy consumption is saved compared with the Original. In our DPT, the processor scales down the supply voltage for a certain period of time when it detects a throughput imbalance between the processor and main memory.

DPT tries to balance the throughput of the processor with that of the main memory by setting the clock frequency and supply voltage of the processor. When the memory performance dominates the execution time, the throughput of the processor is balanced by lowering its frequency. Because the throughput balance changes within a program, the execution of a program is divided into time intervals and the frequency and voltage are adjusted at every interval. DPT is superior to conventional cache-miss-driven methods in the following aspects.

- Energy consumption is further reduced by setting the clock frequency and supply voltage at the balanced point.
- Because voltage switching is less frequent than cache-miss-driven methods, the performance and energy overheads are suppressed.

As for the first aspect, Ishihara and Yasuura [1] proved that total energy consumption is minimized if the processor uses a single supply voltage which fits the execution time just with the given time constraint. This result is extended to our situation. By selecting a clock frequency and supply voltage from several setting points at a balanced point, less energy is consumed compared with switching between high-Vdd and low-Vdd. Suppose the high-Vdd and low-Vdd settings are 1.6 V with 1.8 GHz clock frequency and 0.8 V with 0.6 GHz clock frequency, respectively, as shown in Fig. 1. Then, the energy consumption for the execution related to one cache-miss resolution becomes $2 \times 1.6^2 \times \alpha$ in Original, whereas is is $(1 \times 1.6^2 + 3 \times 0.8^2) \times \alpha$ in Miss-Driven. Consequently, 13% of the energy is reduced in Miss-Driven compared with Original. In DPT, if the balanced Vdd setting is 1.0 V with 0.9 GHz clock frequency, the energy consumption is about $(4 \times 1.0^2) \times \alpha$. Therefore, the energy consumption is further reduced by 11% compared with Miss-Driven without performance penalty.

As for the second aspect, the supply voltage transition is less frequent in DPT than in Miss-Driven. Because performance and energy penalties for switching the clock frequency and supply voltage are not negligible, DVS methods should be carefully designed to take into account these penalties. DPT uses a balanced voltage setting for a certain period of time, whereas Miss-Driven can change the supply voltage at every cache miss or cache-miss resolve. Therefore, DPT is robust with respect to these penalties. The circuit issues for these penalty are discussed in Section 3.3.

3.2   Algorithm of DPT

The DPT algorithm consists of the following three parts: (1) detection of a performance imbalance, (2) prediction of a performance imbalance, (3) adjustment of the supply voltage and clock frequency. The algorithm is based on an interval-based approach, as illustrated in Fig. 2. In this section, we describe how to detect the throughput imbalance between computation and data transfer, how to predict it for future execution, and how to adjust the supply voltage and clock frequency for each time interval. We assume the following memory hierarchy in our explanation.

- The processor has L1 data and instruction caches and an L2 unified cache.
- All the caches are on the processor chip.
- The processor supports non-blocking caches.

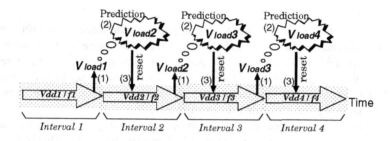

Fig. 2. Interval-based approach

Detecting a Performance Imbalance. The simplest way to estimate the activity of the processor is to use the IPC, as in Pipeline Balancing [5]. When frequent cache misses occur, the IPC of the processor decreases. Therefore, the load of the main memory or the load of the data transfer is estimated as high if the IPC is low. However, the IPC depends on not only on the load of the main memory but also on the instruction level parallelism (ILP) of a program. For example, if the IPC is limited by the dependencies of non-memory-access instructions, the IPC is quite low but the processor is still busy.

Fig. 3(a) shows the relationship between performance degradation and IPC when the frequency of the processor is lowered from 1.6 GHz to 1.4 GHz for every 50000 committed instructions in all programs of the SPEC CPU2000 benchmark suite. The evaluation environment is described in Section 4.1. As seen in this figure, performance degradation is not closely correlated to the IPC. Performance degradation is observed even in low IPC. Therefore, IPC does not always reflect the performance imbalance between the processor and main memory.

Instead of the IPC, we propose a new method to detect the performance imbalance between the processor and main memory. The performance balance between the processor and main memory is estimated based on cache-miss information or more specifically, the number of on-going cache-miss requests. In non-blocking caches, there exist several on-going cache-miss requests at a moment, and they are handled by memory independently of processor execution. In this situation, the number of existing cache-miss requests can be considered as the load of data transfer from and to main memory. Then,

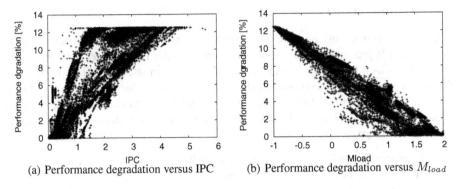

(a) Performance degradation versus IPC   (b) Performance degradation versus M_{load}

Fig. 3. Performance degradation

if multiple cache-miss requests always exist, memory performance is not sufficient for processor performance. This is the imbalance between the processor and memory performance to be tackled by our DPT. Thus, performance imbalance is detected based on the number of existing cache-miss requests. Since main memory accesses are invoked by L2 cache misses in the assumed memory hierarchy, the performance imbalance is detected by watching the L2 cache-miss information.

In the detection, we introduce a status register named Reg_{L2m}, which keeps track of the number of existing cache-miss requests including write-back. The value of Reg_{L2m} is incremented if an L2 cache miss or an L2 cache write-back occurs. When one of these requests is completed, the value is decremented.

In addition to Reg_{L2m}, three counters named $Cnt0$, $Cnt1$, and $Cnt2$ are introduced. In every cycle, the counters of $Cnt0$, $Cnt1$, and $Cnt2$ are incremented when the value of Reg_{L2m} is 0, 1, and more than 2, respectively. By referencing these counters, the distribution of the existence of the number of cache miss requests can be calculated. Using these counters, we quantify the load factor of data transfer (M_{load}) using the following equation.

$$M_{load} = (w_2 \times Cnt2) + (w_1 \times Cnt1) + (w_0 \times Cnt0) \tag{1}$$

Here, w_n indicates the weight value for each Cnt_n.

Fig. 3(b) presents the relationship between performance degradation and average M_{load} per cycle using the same measurement as in Fig. 3(a). We assume that the weight values of w_2, w_1, and w_0 are 2, 1 and -1, respectively. As seen from the figure, performance degradation is almost linearly correlated to M_{load}. This indicates that M_{load} well reflects the performance imbalance between the processor and main memory.

Predicting Performance Imbalance. The performance imbalance between computation and data transfer is predicted by using M_{load}. Let T_{itvl} be the time interval for changing the supply voltage and clock frequency. We calculate M_{load} for every T_{itvl} period.

The current value of M_{load} (in other words, the load factor of the data transfer in the current T_{itvl} period) is used for predicting the next period of the load of the data transfer. The setting for the next interval is determined by this current load value. If M_{load} is high, the load of the data transfer in the next T_{itvl} period is predicted as high. If the value is low, the load in the next period is predicted as low.

Note that the three counters, $Cnt0$, $Cnt1$, and $Cnt2$, are reset at the start point of every time interval.

```
for each cycle {
    if (Reg_L2m == 0) Cnt0++;
    elseif (Reg_L2m == 1) Cnt1++;
    else Cnt2++;
    /* for every T_itvl */
    if ((CycleCount % T_itvl) == 0){
        M_load = (w_2 × Cnt2) + (w_1 × Cnt1) + (w_0 × Cnt0)
        if (M_load > Th_u)
            DownVoltage();
        elseif (M_load < Th_l)
            UpVoltage();
        else
            /* UnchangingVoltage */;
        Cnt0 = Cnt1 = Cnt2 = 0;
    }
    CycleCount + +;
}
```

Fig. 4. Algorithm of DPT

Adjusting the Supply Voltage and Clock Frequency. By using the predicted load of the data transfer, the processor supply voltage and clock frequency are adjusted at each start point of the time intervals. We introduce two threshold values, Th_u and Th_l, which indicate the upper threshold and lower threshold, respectively. The two threshold values are used to avoid thrashing between two voltage settings. At every starting point of T_{itvl}, the voltage setting is raised by one level if M_{load} exceeds Th_u and is lowered by one level if M_{load} is below Th_l. The setting is not changed if the value is between Th_u and Th_l.

We summarize the entire DPT algorithm in Fig. 4.

3.3   Circuit Issues

First, we discuss the additional hardware to realize the DPT algorithm. DPT requires only one register (Reg_{L2m}), three counters ($Cnt0$, $Cnt1$, and $Cnt2$), two adders and three multipliers for calculating expression (1). Here, if simple values are selected as the weight values (2, 1, and -1 are used in the evaluation), a simple shifter or nothing is required for the multiplication. Comparators between M_{load} and Th_u/Th_l are also required but a small number of bits are enough. Thus, DPT implementation does not affect power consumption.

Next, we consider the circuit issues required for switching the clock frequency and supply voltage. Conventional processors with DVS use a dynamic DC-DC converter to change the voltage level. One drawback of a DC-DC converter is that it requires quite a long time for voltage ramping. To suppress this penalty on performance, VSV [8], which is one of the L2-miss-driven techniques, adopts dual power supply networks instead of using the DC-DC converter. However, dual power supply networks have two disadvantages, area overhead and a restriction on voltage choice (only two voltage alternatives are available). On the other hand, because voltage switching is less frequent in DPT, it does not lead to performance degradation even if a DC-DC converter is adopted. Thus, we use a DC-DC converter in DPT. Therefore, DPT is free from the problems of area overhead and the restriction on voltage choice.

As for the energy overhead, a varying supply voltage changes the amount of charge held in the CMOS circuits, as stated in [10]. Because RAM structures contain a number of cells, the energy overhead due to charging/discharging the cells is too large to be amortized. Therefore, VSV does not scale the supply voltage of large RAM structures such as the cache or register file. On the other hand, in DPT, we also scale the supply voltage for large RAM structures. Because voltage switching is less frequent and the difference of voltage levels before and after switching is small, the energy overhead is small even if the supply voltage of all the RAMs changes. Thus, DPT can save power consumption for such RAM accesses.

4   Experimental Setup

4.1   Experimental Methodology

Power consumption and the performance of DPT is evaluated and compared with the original non-DVS model (called original here) and miss-driven DVS model (called miss-driven). The effect of the performance imbalance detection method is also evaluated. We use the SimpleScalar Tool Set [15] for our base simulation environment. Since an accurate simulation of the memory hierarchy is required in the evaluation, we use SimpleScalar augmented with a memory hierarchy extension (SimpleScalar with Memory Extension) [16]. For estimating power consumption, we incorporate the Wattch [17] extension into the environment.

The miss-driven is based on VSV [8] in this evaluation. The supply voltage is changed from high-Vdd to low-Vdd when an L2 cache miss occurs and it is switched back to high-Vdd when all the misses return. The ILP monitoring mechanism is not supported in this evaluation. To avoid a large performance penalty and energy overhead in the voltage transition, it is assumed that a dual power supply network is adopted and high-Vdd is always used for the large RAM structures (caches and register file), as suggested in [8].

We use all the programs from the SPEC CPU2000 benchmark suite using the *ref* input set. The programs are compiled for Alpha instruction set architecture (ISA). We fast-forwarded two billion instructions and simulated 500 million instructions.

4.2 Assumption

Table 2 shows the assumptions of the processor configuration for the evaluation. The supply voltage and the clock frequency used here follow those of the Intel Pentium M processor shown in Table 1. In DPT, the supply voltage and clock frequency were selected from six setting points of Table 1. For miss-driven, the highest and lowest settings of Table 1 were used for high-Vdd and low-Vdd respectively.

The following values are used for the parameters of the DPT algorithm described in Section 3.2.

- T_{itvl}: 100000 cycle
- Th_u: 130000, Th_l: 100000
- $w_2 = 2, w_1 = 1, w_0 = -1$

We incorporate a performance penalty and energy overhead for the frequency/voltage transition into the evaluation. According to the data-sheet of the Intel Pentium M processor[18], the processor core is unavailable for up to $10\mu s$ during frequency ramping. We pessimistically assume $20\mu s$ of penalty for a frequency and voltage transition in DPT. The energy overhead of a transition in DPT is assumed to be $20\mu J$ with consideration of both the charge/discharge of the CMOS circuits and DC-DC converter. For the miss-driven DVS, we optimistically assume that the performance penalty and

Table 1. The combinations of supply voltage and clock frequency on an Intel Pentium M

Processor Clock	1.6 GHz	1.4 GHz	1.2 GHz	1.0 GHz	800 MHz	600 MHz
FSB Clock	400 MHz	400 MHz	400 MHz	400 MHz	400 MHz	400 MHz
Memory Bus Clock	266 MHz	266 MHz	266 MHz	266 MHz	266 MHz	266 MHz
Processor Core Vdd	1.484 V	1.420 V	1.276 V	1.164 V	1.036 V	0.956 V

Table 2. Processor configuration

Fetch width	8
Branch prediction	gshare 4K entry
BTB	1024 sets, 4-way
Mis-prediction penalty	3 cycles
RUU size	128
LSQ size	64
Functional units	Int: 6 ALU, 2 mult/div, FP: 6 ALU 4 mult/div Load/Store: 2 memory ports
L1 instruction-cache	32 KB, 32 B line, 2-way 1 cycle latency
L1 data-cache	32 KB, 32 B line, 2-way 2 cycle latency
L2 cache	512 KB, 64 B line, 8-way 10 cycle latency
Bus width	8 B
Main Memory latency	50 ns

energy overhead for a frequency/voltage transition are $10ns$ and $22.3nJ$ respectively, which is quite favorable for a miss-driven DVS.

5   Evaluation Result

5.1   Comparison Between Miss-Driven DVS and DPT

Fig. 5 shows the relative execution time and relative energy consumption of the DPT and miss-driven normalized to the original non-DVS model. The execution time shown in Fig. 5(a) is decomposed into seven parts. Six parts correspond to the time executed in the six different processor clock frequencies, and the other part represents the time penalty required for changing the supply voltage level. Programs are sorted in the order of the L2 cache-miss rate (programs with higher L2 miss rates are located on the left side). In Fig. 5(b), each bar is decomposed into two parts: the energy for program execution and the additional energy overhead for voltage transitions. Note that the results for programs whose L2 cache-miss rate is smaller than 0.1% are not shown in Fig. 5 because the performance and energy consumption of DPT and miss-driven for such

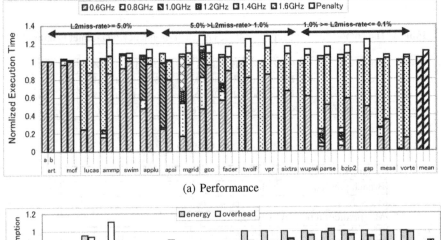

(a) Performance

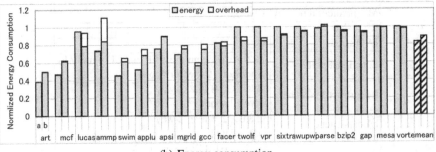

(b) Energy consumption

Fig. 5. Results on SPEC2000 programs. In the figure, labels of "a" and "b" represent DPT and miss-driven, respectively.

programs are close to those of the original processor. The mean bars shown on the far right side of the figure indicate the mean value across all of the SPEC2000 programs.

First of all, we discuss the impact on performance. By averaging all the programs, 3.4% and 11% performance degradation is observed in DPT and miss-driven, respectively. In most of the programs, DPT outperforms miss-driven in performance. This result is derived from the following two observations. First, the execution time of DPT, excluding the penalty, is shorter than that of miss-driven except for mcf, apsi and gcc. Second, the penalty of DPT is smaller than miss-driven except for gcc. The reason for the first observation is that the miss-driven processor uses only two kinds of clock frequencies (1.6 GHz and 0.6 GHz), whereas the DPT processor uses six kinds of frequencies. The miss-driven often selects the lowest clock frequency, but the frequency is sometimes too low and the instructions in critical passes are delayed. On the other hand, because DPT successfully sets the clock frequency stably at the balanced point between the processor and main memory for a time interval, performance degradation is mitigated. This reasoning is supported by the fact that intermediate frequencies are often selected in DPT, as shown in Fig. 5(a).

The reason for the second observation is the frequency of the voltage transition. Table 3 presents the number of voltage transitions for both methods. Because the number of voltage transitions in DPT is quite small, the penalty is suppressed even though the time loss for one voltage switching in DPT is four orders of magnitude longer than that of miss-driven. This suggests that the voltage transition should be triggered not by specific events but by capturing the execution status for a certain period.

Next, we discuss the results of energy consumption. Both miss-driven and DPT save a significant amount of energy in the programs with a high L2 cache-miss rate (greater than 5%) except for lucas. Because the processor wastes a lot of time in waiting for data from the main memory in these programs, energy consumption is reduced with little performance degradation by lowering the supply voltage. For example, DPT saves 61% and 53% energy with 1% and 3.5% performance penalty in art and mcf, respectively.

Comparing the energy consumption of DPT with that of miss-driven, the energy consumption is further reduced in DPT in most of the applications with a high cache-

Table 3. The number of voltage transitions

	art	mcf	lucas	ammp	swim	applu	apsi
DPT	5	1290	27	755	639	125	54
miss-driven	575373	1708023	14601666	23569591	4422766	8086547	574946

	mgrid	gcc	facerec	twolf	vpr	sixtrack	wupwise
DPT	72	6225	652	0	2	4	0
miss-driven	4606348	9082303	5528369	6883742	5372244	1569380	973562

	parser	bzip2	gap	mesa	vortex	crafty	gzip
DPT	514	475	0	22	36	0	0
miss-driven	2781574	2134300	1474962	416284	617372	208472	98486

	galgel	equake	eon	fma3d	perlbmk	-	-
DPT	0	0	0	0	0	-	-
miss-driven	39294	26274	2400	844	204	-	-

miss rate. This indicates that energy consumption is further minimized by setting the supply voltage at the balanced point than by frequently switching the high and low voltages. For some programs (twolf, vpr, sixtrack, wupwise, bzip2, and gap), miss-driven seems to save more energy than DPT. However, this is achieved at the sacrifice of performance. As for the energy overhead due to voltage transitions, because the number of transitions in DPT is much smaller than that that in miss-driven, the overhead is much reduced. On average, the energy saving for DPT reaches 17% while that of miss-driven is 11%.

Because DPT is superior to miss-driven both in performance and energy, DPT is more efficient than miss-driven for reducing energy consumption with only a little performance degradation. Although the performance degradation in miss-driven can be mitigated by raising the voltage level of low-Vdd or by applying the ILP monitoring mechanism suggested in [8], it always leads to the loss of opportunities for energy saving. Thus, these evaluation results indicate that balancing throughput between the processor and main memory is a very promising way to low-power computing.

5.2 Effect of Performance Imbalance Detection Method

To examine the effect of the performance imbalance detection method, we compare DPT with the IPC-based DVS method (hereafter denoted as IPC-based). The algorithm of IPC-based is the same as DPT except for performance imbalance detection; that is, IPC is used to estimate the imbalance.

Fig. 6 and Fig. 7 present the performance degradation and energy saving of DPT and IPC-based compared with the original non-DVS processor on average across all the SPEC2000 programs. Three kinds of upper and lower threshold pairs are evaluated for

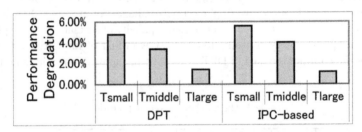

Fig. 6. Performance degradation of DPT and IPC-based

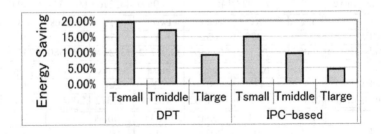

Fig. 7. Energy saving of DPT and IPC-based

each detection method. The upper and lower threshold values (upper–lower) of $Tsmall$, $Tmiddle$, and $Tlarge$ are 110000–70000, 130000–100000, and 150000–130000, respectively, for DPT and 0.7–0.5, 0.5–0.3, and 0.3–0.1, respectively, for IPC-based.

Comparing DPT with IPC-based in the threshold of $Tsmall$ or $Tmiddle$, DPT saves more energy than IPC-based, even though the performance degradation of DPT is less than IPC-based. As for the case of $Tlarge$, though the performance degradation of DPT is almost the same as IPC-based, the energy saving of DPT is twice as much as that of IPC-based. Because M_{load} used in DPT well reflects the load of the main memory as shown in Section 3.2, a large amount of energy is saved with a small performance penalty compared with IPC-based. Therefore, it is concluded that the proposed performance imbalance detection method is superior to the IPC-based detection method.

6   Concluding Remarks

This paper proposed a hardware-based DVS technique called Dynamic Processor Throttling (DPT) for power efficient computations. DPT reduces the power/energy consumption by detecting the performance imbalance between the processor and main memory and adjusts the processor supply voltage and clock frequency to redress the imbalance. A method to detect the imbalance was also proposed.

We evaluated the effect of power saving and the impact on performance by using the SimpleScalar framework and SPEC CPU2000 benchmark programs. The evaluation results revealed that the proposed DPT can reduce more energy with less performance degradation compared with the miss-driven approach. In DPT, 17% of the energy consumption is reduced with 3.4% performance degradation. We also evaluated the effect of the performance imbalance detection method. It is revealed that DPT can estimate the load of the main memory more accurately than the IPC-based detection method. From these evaluation results, it is concluded that DPT is very effective for low-power computing.

Acknowledgment

This work is partly supported by STARC (Semiconductor Technology Academic Research Center) and the Ministry of Education, Culture, Sports, Science and Technology, Grant-in-Aid (No. 14380136).

References

1. T. Ishihara and H. Yasuura: Voltage Scheduling Problem for Dynamically Variable Voltage Processors. In Proc. ISLPED1998, pp.197–201, Aug. 1998.
2. G. Qu: What is the Limit of Energy Saving by Dynamic Voltage Scaling. In Proc. ICCAD 2001, pp.560–563, Nov. 2001.
3. D. Shin and J. Kim: A Profile-Based Energy-Efficient Intra-Task Voltage Scheduling Algorithm for Hard Real-Time Applications. In Proc. ISLPED2001, pp.271–274, Aug. 2001.

4. R. Sasanka, C.J. Hughes, and S.V. Adve: Joint Local and Global Hardware Adaptations for Energy. In Proc. ASPLOS X, pp.144–155, Oct. 2002.
5. R.I Bahar and S. Manne: Power and Energy Reduction via Pipeline Balancing. In Proc. 28th ISCA, pp.218–229, Jul. 2001.
6. C.-H. Hsu, et al.:Compiler-Directed Dynamic Frequency and Voltage Scheduling. In Proc. Workshop on PACS, Nov. 2000.
7. D. Marculescu: On the Use of Microarchitecture-Driven Dynamic Voltage Scaling. In Proc. of Workshop on Complexity-Effective Design, June 2000.
8. H. Li, C-Y. Cher, T.N. Vijaykumar, and K. Roy: VSV: L2-Miss-Driven Variable Supply-Voltage Scaling for Low Power. In Proc. 36th Micro, pp.19–28, Dec. 2003.
9. P.S-Marbell, et al.: A Hardware Architcture for Dynamic Performance and Energy Adaptation. In Proc. Workshop on PACS, Feb. 2002.
10. T.D. Burd and R.W. Brodersen: Design Issues for Dynamic Voltage Scaling. In Proc. ISLPED2000, pp.9–14, Aug. 1998.
11. H. Saputra, et al.: Energy-Conscious Compilation Based on Voltage Scaling. In Proc. LCTES/SCOPES 2002, pp.2–11, June 2002.
12. G. Semeraro, et al.: Energy-Efficient Processor Design Using Multiple Clock Domains with Dynamic Voltage and Frequency Scaling. In Proc. the 8th HPCA, pp.29–40, Feb. 2002.
13. G. Semeraro, et al.: Dynamic Frequency and Voltage Control for a Multiple Clock Domain Microarchitecture. In Proc. 35th Micro, pp.356–367, Dec. 2002.
14. G. Magklis, et al.: Profile-based Dynamic Voltage and Frequency Scaling for a Multiple Clock Domain Microprocessor. In Proc. the 30th ISCA, pp.14–27, June 2003.
15. T. Austin, et al.,: SimpleScalar: An Infrastructure for Computer System Modeling. IEEE Computer, Vol. 35, No. 2, pp.59–67, Feb. 2002.
16. D. Burger, et al.: Memory Hierarchy Extensions to the SimpleScalar Tool Set. Technical Report TR99-25, Department of Computer Science, University of Texas at Austin, April 1999.
17. D. Brooks, et al.: Wattch: A Framework for Architectural-Level Power Analysis and Optimizations. In Proc. 27th ISCA, pp.83–94, June 2000.
18. Intel: Intel Pentium M Processor Datasheet. June 2003.

Effective Dynamic Voltage Scaling Through CPU-Boundedness Detection*

Chung-Hsing Hsu and Wu-Chun Feng

Los Alamos National Laboratory,
Los Alamos, U.S.A.
{chunghsu, feng}@lanl.gov

Abstract. Dynamic voltage scaling (DVS) allows a program to execute at a non-peak CPU frequency in order to reduce CPU power, and hence, energy consumption; however, it is oftentimes done at the expense of performance degradation. For a program whose execution time is bounded by peripherals' performance rather than the CPU speed, applying DVS to the program will result in negligible performance penalty. Unfortunately, existing DVS-based power-management algorithms are *conservative* in the sense that they overly exaggerate the impact that the CPU speed has on the execution time. We propose a new DVS algorithm that detects the CPU-boundedness of a program on the fly (via a regression method on the past MIPS rate) and then adjusts the CPU frequency accordingly. To illustrate its effectiveness, we compare our algorithm with other DVS algorithms on real systems via physical measurements.

1 Introduction

Dynamic voltage and frequency scaling (DVS) is a mechanism whereby software can dynamically adjust CPU voltage and frequency. This mechanism allows systems to address the problem of ever-increasing CPU power dissipation and energy consumption, as they are both quadratically proportional to the CPU voltage. However, reducing the CPU voltage may also require the CPU frequency to be reduced and results in degraded CPU performance with respect to execution time. In other words, DVS trades off performance for power and energy reduction.

The performance loss due to running at a lower CPU frequency raises several issues. First, a user who pays to upgrade his/her computer system does not want to experience performance degradation. Second, running programs at a low CPU frequency may end up increasing total system energy usage [1,2,3]. In order to control (or constrain) the performance loss effectively, a model that relates performance to the CPU frequency is essential for any DVS-based power-management algorithm (shortened as DVS algorithm hereafter).

A typical model used by many DVS algorithms predicts that the execution time will double if the CPU speed is cut in half. Unfortunately, this model

---

* Available as technical report LA-UR-04-7195.

B. Falsafi and T.N. Vijaykumar (Eds.): PACS 2004, LNCS 3471, pp. 135–149, 2005.

overly exaggerates the impact that the CPU speed has on the execution time. It is only in the worst case that the execution time doubles when the CPU speed is halved; in general, the actual execution time is less than double. For example, in programs with a high cache miss ratio, performance can be limited by memory bandwidth rather than CPU speed. Since memory performance is not affected by a change in CPU speed, increasing or decreasing the CPU frequency will have little effect on the performance of these programs. We call this phenomenon — *sublinear performance slowdown*. Consequently, researchers have been trying to exploit this program behavior in order to achieve better power and energy reduction [4,5,6,7].

One common technique to exploit the sublinear performance slowdown decomposes program workload into regions based on their CPU-boundedness. The decomposition can be done statically using profiling information [4] or dynamically through an auxiliary circuit [5] or through a built-in performance monitoring unit (PMU) [6,7]. In this paper, we propose a new PMU-assisted, on-line DVS algorithm called *β-adaptation* that provides fine-grained, tight control over performance loss and takes advantage of sublinear performance slowdown. This new *β-adaptation* algorithm is based on an extension of the theoretical work developed by Yao et al. [8] and by Ishihara and Yasuura [9]. Via physical measurements, we will demonstrate the effectiveness of the *β-adaptation* algorithm when compared to several existing DVS algorithms for a number of applications.

The rest of the paper is organized as follows. Section 2 characterizes how current DVS algorithms relate performance to CPU frequency. With this characterization as a backdrop, we present a new DVS algorithm (Section 3) along with its theoretical foundation (Section 4). Then, Section 5 describes the experimental set-up, the implemented DVS algorithms, and the experimental results. Finally, Section 6 concludes and presents some future directions.

2   Related Work

There have been some attempts to exploit the sublinear performance slowdown (where increasing or decreasing the CPU frequency will have little effect on the performance of a program) to achieve more power and energy reduction. For example, Li et al. [5] propose to set the CPU to a low speed whenever an L2 cache miss occurs, whereas Hsu and Kremer [4] use off-line profiling to identify memory-bound program regions. The former approach requires an auxiliary circuit, and the latter approach needs source code and compiler support. These requirements make their approaches more difficult to implement in practice.

Another approach is to use built-in performance monitoring unit (PMU) to assist in the on-line detection of sublinear performance slowdown. Our work and Choi et al.'s recent work [6,7] belong to this category. Both use a regression method and PMU support to perform the on-line construction of a simple performance-prediction model so as to capture the degree of CPU-boundedness. In general, the design of PMU-assisted on-line DVS algorithms is not an easy task. First, the PMU is notorious for its incomplete set of event counting and inconsistency across generations of the CPU. Second, the correlation of event counts to power and performance is not yet clear. Hence, for now, a PMU-assisted, on-line, DVS algorithm ought to minimize its dependency on event counts and rely as much as possible on those event counts that are consistent across CPU generations.

Our work differs from Choi et al.'s work in the definition of CPU-boundedness, and thus, the detection mechanism. Choi et al.'s work is based on the ratio of the on-chip computation time to the off-chip access time. In contrast, our algorithm defines CPU-boundedness as the fraction of program workload that is CPU-bound. Because of the different definitions, the set of events monitored by the PMU for each algorithm is different. In Section 5.5, we argue that our DVS algorithm is equally effective but has a simpler implementation. Moreover, we provide a theoretical foundation of why our DVS algorithm is effective in achieving energy optimality. We believe that the same theoretical result can be applied to their work as well.

3   β-Adaptation: A New DVS Algorithm

Here we describe a new, interval-based, PMU-assisted, DVS algorithm that provides fine-grained, tight control over performance loss as well as exploits the sublinear performance scaling in memory-bound and I/O-bound programs. The theoretically-based heuristic algorithm is based on an extension of the theoretical work developed by [8] and [9] (details in Section 4):

> If the CPU power draw is a convex function of the CPU frequency, then for any program whose performance is an *affine* function of the CPU frequency, running at a constant CPU speed and meeting the deadline just in time will minimize the energy usage of executing the program. If the desired CPU frequency is not directly supported, the two immediately-neighboring CPU frequencies can be used to emulate the desired CPU frequency and result in an energy-optimal DVS schedule.

To account for the sublinear performance slowdown, the following model that relates performance to the CPU frequency is often used [6,7,10]:

$$T(f) = W_{cpu} \cdot \frac{1}{f} + T_{mem} \tag{1}$$

The total execution time $T(f)$ at frequency f is decomposed into two parts. The first part models on-chip workload in terms of CPU cycles. Its value is affected by the CPU speed change. The second part models the time due to off-chip accesses and is invariant to changes in the CPU speed. Note that this breakdown of the total execution time is inexact when the target processor supports out-of-order execution because on-chip execution may overlap with off-chip accesses [11]. However, in practice, the error tends to be quite small [6,7].

The model $T(f)$ treats program performance as an affine function of the CPU frequency f and thus allows us to apply the aforementioned theoretical result.

We simply execute a program at CPU frequency f^* such that $D = T(f^*)$ where D is the deadline of the program. However, there are two challenges in using the theorem this way. First, in many cases there is no consensus on how to assign a deadline to a program, e.g., scientific computation. Second, to use $T(f)$, we need to know the values of the coefficients, W_{cpu} and T_{mem}. These coefficients are oftentimes determined by the hardware platform, program source code, and data input. Thus, calculating these coefficients statically is very difficult.

We address these challenges by defining a deadline as the relative performance slowdown and by estimating the model's coefficients on the fly (without any off-line profiling nor compiler support). The relative performance slowdown δ

$$\delta = \frac{T(f)}{T(f_{max})} - 1 \tag{2}$$

where f_{max} is the peak CPU frequency, has been used in previous work [6,7,11]. It is widely accepted in programs that are difficult to assign deadlines in terms of absolute execution time. It also carries more timing requirement information than CPU utilization and IPC rate. Providing this user-tunable parameter δ in our DVS algorithm allows fine-grained, tight control over performance loss.

To estimate the coefficients more efficiently, we first re-formulate the original two-coefficient model in Equation (1) as a single-coefficient model:

$$\frac{T(f)}{T(f_{max})} = \beta \cdot \frac{f_{max}}{f} + (1 - \beta) \tag{3}$$

with

$$\beta = \frac{W_{cpu}}{W_{cpu} + T_{mem} \cdot f_{max}} \tag{4}$$

The coefficient β is, by definition, a value between 0 and 1. It was introduced by one of the authors in [4] to quantify, for a program, the performance impact to the CPU speed change. The metric represents the fraction of the program workload that scales linearly with the CPU frequency. If a program has $\beta = 1$, it means the execution time of the program will double when the CPU speed is halved. In contrast, a program with $\beta \approx 0$ will have its execution time remained the same even running at the slowest CPU speed.

The coefficient β is computed at run time using a regression method on the past MIPS rates reported from the PMU. Specifically, our DVS algorithm keeps track of the average MIPS rate for each executed CPU frequency and applies the least-square fitting at each interval to dynamically re-compute the new β value:

$$\beta = \frac{\sum_i (\frac{f_{max}}{f_i} - 1)(\frac{\text{mips}(f_{max})}{\text{mips}(f_i)} - 1)}{\sum_i (\frac{f_{max}}{f_i} - 1)^2} \tag{5}$$

where $\text{mips}(f)$ is the average MIPS rate for CPU frequency f. Note that our mechanism assumes a constant number of total instructions in a program, regardless of the running CPU frequency. This assumption has been verified through

```
For every I seconds, do the following:
```

1. Use Equation (5) to compute β.
2. Compute frequency f^*.

$$f^* = \max\left(f_{min}, \frac{f_{max}}{1 + \delta/\beta}\right)$$

3. Figure out f_j and f_{j+1}.

$$f_j \leq f^* < f_{j+1}$$

4. Compute the ratio r.

$$r = \frac{1/f^* - 1/f_{j+1}}{1/f_j - 1/f_{j+1}}$$

5. Run $r \cdot I$ seconds at f_j.
6. Run $(1 - r) \cdot I$ seconds at f_{j+1}.
7. Update $\mathrm{mips}(f_j)$ and $\mathrm{mips}(f_{j+1})$.

Fig. 1. Algorithm β-*adaptation*. Parameter δ is the relative performance slowdown and parameter I is the length of an interval in seconds.

extensive experiments. In practice, the value of β converges very quickly for the benchmarks we tested.

The rest of the algorithm simply applies the theoretical result to compute the desired CPU frequency f^* for each interval, once the coefficient β is updated, plus some bookkeeping on $\mathrm{mips}(f)$. The derivation of f^* comes by equating Equation (2) with Equation (3). Figure 1 outlines the entire algorithm.

4   Theoretical Foundation

In the previous section, we claim a theoretical result for energy-optimal DVS scheduling which extends both Yao et al.'s work in [8] and Ishihara and Yasuura's work in [9]. In this section we provide evidence to support our claim.

The energy-optimal DVS scheduling problem considered here is taken from [4]. That previous work only provides a problem formulation. In this paper we provide a theorem that characterizes the energy-optimal DVS schedule for the problem. The theorem is also closely related to previous work such as Miyoshi et al.'s "critical power slope" [2].

A DVS system is assumed to export n settings $\{(f_i, P_i)\}$, where P_i is the CPU power dissipation (in watts) at CPU frequency f_i. Without loss of generality, we assume $0 < f_{min} = f_1 < \cdots < f_n = f_{max}$. We also denote the total execution time of a program running at setting i as T_i. Finally, to facilitate discussion, we define $E_i = P_i \cdot T_i$, where E_i is the energy consumption (in joules) when running for T_i seconds at CPU frequency f_i.

The DVS scheduling problem is formulated as follows: Given a program and a deadline D (in seconds), find a DVS schedule $(t_1^*, \cdots, t_n^*)$ such that if the program is executed for t_i^* seconds at setting i, the total energy usage E is minimized, the deadline D is met, and the required work is completed. Mathematically speaking,

$$t^* = \arg\min\{E = \sum_i P_i \cdot t_i : \sum_i t_i \leq D, \sum_i t_i/T_i = 1, t_i \geq 0\} \qquad (6)$$

To simplify the discussion of the theorem, we handle a few corner cases first. First, the condition $D \geq \min_i T_i$ has to be satisfied so that the problem is feasible. Second, if the condition $D \geq \max_i T_i$ is satisfied, the problem becomes the classical fractional Knapsack problem [12]. In this case, the energy-optimal DVS schedule will execute the entire program at setting i^* where $i^* = \arg_i \min\{E_i\}$. For the case of $T_1 = \cdots = T_n$, the above DVS schedule is also energy-optimal. What is left is the case $\min_i T_i < D < \max_i T_i$, which we assume to be true for the following theorem.

Theorem 1. *If*

$$T_i = \frac{c_1}{f_i} + c_0, \ c_1 \neq 0$$

and

$$\frac{P_1 - 0}{f_1 - 0} \leq \frac{P_2 - P_1}{f_2 - f_1} \leq \frac{P_3 - P_2}{f_3 - f_2} \leq \cdots \leq \frac{P_n - P_{n-1}}{f_n - f_{n-1}}$$

then

$$t_i^* = \begin{cases} \frac{1/f^* - 1/f_{j+1}}{1/f_j - 1 - f_{j+1}} \cdot T_j & i = j \\ D - t_j^* & i = j+1 \\ 0 & \text{otherwise} \end{cases}$$

where

$$f_j \leq f^* < f_{j+1}$$

Proof. (See the Appendix).

Theorem 1 says that for any program whose execution time is an *affine* function of the CPU frequency, if the DVS settings in a CPU are *well-assigned* (explained below), then we can run the program at a CPU frequency that finishes the execution right at the deadline *and* results in an energy-optimal schedule. If the desired CPU frequency is not directly supported, it can be emulated by the two immediately-neighboring CPU frequencies.

For any DVS-enabled processor whose power draw can be modeled as a convex function of its frequency, the processor's DVS settings are always well-assigned. However, some realistic processors do not have well-assigned DVS settings by default. In these processors, the lowest frequency f_1 can be emulated by the combination of frequency 0 (i.e., the CPU in sleep mode) and the second lowest frequency f_2 with a *lower* power dissipation, i.e., $\frac{P_1 - 0}{f_1 - 0} > \frac{P_2 - P_1}{f_2 - f_1}$. As a result, completing a task *before* its deadline and putting the CPU into sleep mode is more energy-efficient than completing the task at the deadline. This is the phenomenon observed by Miyoshi et al. [2] and motivated them to devise a technique called "critical power slope". The phenomenon can be eliminated by making adjustments to DVS settings so that they become well-assigned.

Finally, Theorem 1 extends the work presented by Yao et al. [8] and by Ishihara and Yasuura [9]. First, both works assume that $c_0 = 0$. Second, Ishihara and Yasuura's work assumes a fixed relationship between f and V in a DVS setting; namely,

$$f = k \cdot (V - V_T)^\alpha / V \tag{7}$$

where k, V_T, α are positive constants. Unfortunately, today's DVS processors may not be able to support such an assumption. This is because these processors only provide a discrete set of CPU frequencies and voltages, whereas the above equation requires a continuous range of CPU frequencies to be supported for a discrete set of voltages. Theorem 1 loosens these assumptions to facilitate DVS algorithms on realistic processors.

5   Experiments

In this section, we describe our experimental environment in which we evaluate and compare algorithm β-adaptation with several other DVS algorithms. We also present an in-depth discussion of the experimental results.

5.1   Experimental Setup

In order to acquire high-fidelity experimental data, we set-up our experiments using physical measurements, as shown in Figure 2(a). The experimental results were collected through a Yokogawa WT210 digital power meter [13]. The power meter continuously samples the instantaneous wattage at every 20 μs. The profiling and tested computer both run the Linux 2.4.18 kernel. All the benchmarks were compiled by GNU compilers with optimization level -O2. All the benchmarks were run to completion; each run took over a minute.

The benchmarks are taken from SPEC's CPU95 benchmark suites. The SPEC benchmarks [14] emphasize the performance of the CPU and memory, but not other computer components such as I/O (disk drives), networking or graphics. We chose to use the SPEC benchmarks because they demonstrate a range of performance sensitivity to the CPU frequency change, i.e., they have a wide range of β values [4]. The experimental data are collected by running these SPEC benchmarks with the reference data input.

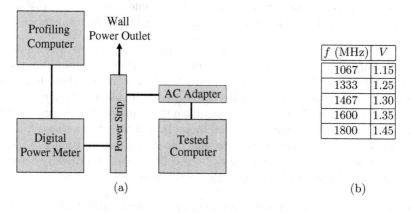

f (MHz)	V
1067	1.15
1333	1.25
1467	1.30
1600	1.35
1800	1.45

(a)   (b)

Fig. 2. The experimental setup

The hardware platform in our experiments is an HP NX9005 notebook computer. This computer includes a mobile AMD Athlon XP 2200+ processor, 256-MB DDR SDRAM, 266-MHz front-side bus, a 30-GB hard disk, and a 15-inch TFT LCD display. The mobile AMD Athlon XP processor has been used in Sun's Fire B100x blade servers [15]. It has a total of 384-KB cache space. The processor exports two registers that the software can write the target frequency and voltage values into. In our experiments, we restrict the processor to have five settings as shown in Figure 2(b). The transition time from one setting to another is 100 microseconds. During the measurements, the battery was removed, and the monitor was turned off.

Finally, when presenting the experimental results, we associate with each application its β value. Recall that the metric β represents the fraction of the program workload that is very sensitive to the CPU speed change. That is, the higher the β of a program, the more CPU-bound its performance. The β value for each benchmark was derived by profiling total execution times for all settings and then applying a least-squares fit on Equation (3).

5.2 Implemented DVS Algorithms

To evaluate the effectiveness of our DVS algorithm β-adaptation, we have implemented a number of other DVS algorithms. Though we do not claim that the implemented DVS algorithms represent a comprehensive comparison of all existing approaches, we feel that the range is wide enough to evaluate the effectiveness of our algorithm and to gain new insights from the experimental results. The following is a brief description of each algorithm we implemented.

2step: This algorithm assumes dual CPU speeds in the processor and monitors the CPU utilization percentage periodically. If the percentage is higher than a pre-defined threshold, the algorithm will set the CPU to the fast speed; if it is lower than another pre-defined threshold, the algorithm will set the CPU to the low speed. This DVS algorithm is considered to be the best algorithm in Grunwald et al.'s empirical study on several interval-based algorithms using CPU utilization [16]. In our implementation, the two thresholds are 50% and 10% and the two speeds are the maximum and the minimum CPU speeds in the processor.

nqPID: This algorithm was proposed by Varma et al. [17] as a refinement of the *2step* algorithm. Recognizing the similarity of DVS scheduling and a classical control-systems problem, the authors took the equation describing a PID controller (Proportional-Integral-Derivative) and modified it to suit the DVS scheduling problem. This algorithm significantly improved the control over performance loss that the *2step* algorithm lacks. In addition, the authors found out that the algorithm's effectiveness does not depend on careful tuning of parameters, which is a nice feature given that *2step*'s effectiveness is critically dependent on the choice of application-specific threshold values [16].

freq: This algorithm is similar to strategies that reclaim the slack time between the actual processing time and the worst-case execution time (e.g., [18,19]).

Specifically, the algorithm keeps track of the amount of remaining CPU work W_{left} and the amount of remaining time before the deadline T_{left}. The desired CPU frequency f_{new} at each interval is simply

$$f_{new} = \frac{W_{left}}{T_{left}}.$$

The algorithm assumes that the total amount of work in CPU cycles is known a priori, which, in practice, is often unpredictable [1] and not always a constant across frequencies [10].

mips: This algorithm is taken from [20] and represents a DVS strategy guided by an externally specified performance metric. Specifically, the new frequency f_{new} at each interval is computed by

$$f_{new} = f_{prev} \cdot \frac{\text{MIPS}_{target}}{\text{MIPS}_{observed}}$$

where f_{prev} is the frequency for the previous interval, MIPS_{target} is the externally specified performance requirement, and $\text{MIPS}_{observed}$ is the real MIPS rate observed in the previous interval. In our experiments, each benchmark has its own MIPS_{target}, which is derived by measuring the MIPS rate for the entire application and then dividing it by $(1 + \delta)$.

5.3 Experimental Results

Table 1 presents the experimental results for the five interval-based DVS algorithms. When a program is memory-bound or I/O-bound (β close to zero), there is substantial opportunity to reduce CPU energy consumption with negligible performance loss. In contrast, when a program is CPU-bound, there is little opportunity to reduce CPU power and energy within a tight performance-loss bound of 5%. Moreover, none of these five DVS algorithms could produce a DVS schedule that had the exact performance degradation of 5%; the actual performance loss varied from one benchmark to another.

Among the five interval-based DVS algorithms, the *β-adaptation* algorithm outperforms the others. In a sense, it verifies that our mechanism for computing CPU-boundedness on the fly is of low overhead and that the algorithm is effective in providing tight control over performance loss due to DVS as well as exploiting the sublinear performance slowdown for significantly more CPU power and energy savings. Algorithms *mips* and *nqPID* arguably rank second. Algorithm *mips* delivers better control over performance loss for all eight benchmarks that we tested, whereas algorithm *nqPID* performs better with respect to power and energy reduction but at the expense of more substantial performance loss. This is especially obvious for the CPU-bound benchmarks. Algorithms *freq* and *2step* clearly rank last.

So, what have we learned from this experiment? First, the number of instructions is a better metric for specifying the CPU work requirement than the

Table 1. The effectiveness of 5 different DVS algorithms. Each table entry is in the format of *relative-time/relative-energy* with respect to the total execution time and system energy usage when running the application at the highest setting throughout the entire execution.

program	β	2step	nqPID	freq	mips	β-adapt.
swim	0.02	1.00/1.00	1.04/0.70	1.00/0.96	1.00/1.00	1.04/0.61
tomcatv	0.24	1.00/1.00	1.03/0.69	1.00/0.97	1.03/0.83	1.00/0.85
su2cor	0.27	0.99/0.99	1.05/0.70	1.00/0.95	1.01/0.96	1.03/0.85
compress	0.37	1.02/1.02	1.13/0.75	1.02/0.97	1.05/0.92	1.01/0.95
mgrid	0.51	1.00/1.00	1.18/0.77	1.01/0.97	1.00/1.00	1.03/0.89
vortex	0.65	1.01/1.00	1.25/0.81	1.01/0.97	1.07/0.94	1.05/0.90
turb3d	0.79	1.00/1.00	1.29/0.83	1.03/0.97	1.01/1.00	1.05/0.94
go	1.00	1.00/1.00	1.37/0.88	1.02/0.99	0.99/0.99	1.06/0.96

number of CPU cycles. For the benchmarks we tested, we found that the number of instructions tends to remain constant across all settings. In contrast, the number of CPU cycles varies significantly depending on the executed DVS schedule. For example, the swim benchmark, when running at the lowest setting, has only 60% of the CPU execution cycles running at the highest setting. Typically, algorithm *freq* uses the worst-case execution cycles which in our case is the number of CPU cycles at the highest setting. This approach exaggerates the amount of the CPU work to be done and results in less effective energy reduction. This explains why algorithm *mips* performs better than algorithm *freq*.

Second, a large window size of past PMU reports is better than a small window size of past PMU reports. In the experiments we found that the MIPS rate varies significantly from interval to interval, especially for CPU-intensive applications. However, the accumulated MIPS rate converges quickly. Thus, the use of the MIPS rate in a global manner seems to be more effective than the use of the rate in a local manner. This partially explains the effectiveness of algorithm *β-adaptation* compared to algorithm *mips*. One concern, however, for using a large window size is that the DVS algorithm may be less responsive for programs that expose multiple execution phases of varying degrees of CPU-boundedness. For the SPEC benchmarks, which are known to have the aforementioned behavior, this does not seem to be a problem. More details can be found in Section 5.4.

Finally, we confirmed that CPU utilization by itself does not provide enough information about system timing requirements. As a result, the control over performance loss is unsatisfactory. This can be seen from the experimental results of algorithm *2step* and algorithm *nqPID*. Algorithm *2step* does not seem to perform any DVS scheduling. This is because the CPU for SPEC benchmarks is active almost all the time, i.e., its CPU utilization is always full. In this case, there exists no optimal threshold values for *2step* to make it more effective. Algorithm *nqPID* refines algorithm *2step* by removing the threshold mechanism from the end user. While it is more effective than algorithm *2step* in terms of CPU power and energy reduction, the lack of enough information about deadlines makes it impossible to provide tight control over performance loss.

5.4   The Impact of Multiple-Phase Execution Behavior

To better address the impact of multiple-phase programs to the DVS algorithm β-adaptation, we compare it with a profile-based, off-line DVS algorithm called hsu [4]. The algorithm hsu uses PMU-assisted off-line profiling and source code analysis to identify the most energy-profitable region in a program to slow down without causing the performance loss to surpass a pre-defined level. Off-line profiling is performed on a section-by-section basis while the DVS scheduling decisions are made in a global manner, competitively comparing the different sections. This global view of the impact of DVS on different code sections allows more effective DVS scheduling, especially for multiple-phase programs such as the SPEC benchmarks.

Algorithm hsu also uses the relative performance slowdown δ to specify control over performance loss. Thus, it allows us to compare the two algorithms on a fair basis. In the experiments we executed the profile-based algorithm hsu with two different training inputs, denoted as hsu(train) and hsu(ref) respectively. The two sets of training inputs are provided along with the SPEC benchmark codes. Table 2 shows the experimental results of both algorithms for the CFP95 benchmark suite.

We conclude that the effectiveness of algorithm β-adaptation is comparable to that of algorithm hsu. Both algorithms achieve a significant amount of CPU power and energy reduction with tight control over performance loss. It is interesting to note that the two algorithms seem to complement each other. Algorithm β-adaptation performs better in CPU-bound benchmarks from mgrid to fpppp, whereas algorithm hsu performs better in memory-bound benchmarks from swim to hydro2d. We are in the process of investigating the causes for this phenomenon.

As mentioned in Section 2, the effectiveness of profile-based DVS algorithms is highly determined by its training data input. In our experiments, we found

Table 2. The comparison of our new on-line DVS algorithm β-adaptation with an off-line DVS algorithm hsu. Each table entry is in the format of relative-time/relative-energy with respect to the total execution time and system energy usage when running the application at the highest setting throughout the entire execution.

program	β	hsu(train)	hsu(ref)	β-adapt.
swim	0.02	1.01/0.75	1.04/0.59	1.04/0.61
tomcatv	0.24	1.03/0.70	1.06/0.60	1.00/0.85
hydro2d	0.19	1.03/0.75	1.03/0.79	1.02/0.84
su2cor	0.27	1.01/0.88	1.02/0.83	1.03/0.85
applu	0.34	1.03/0.87	1.03/0.87	1.04/0.85
apsi	0.37	1.03/0.85	1.04/0.91	1.05/0.83
mgrid	0.51	1.01/1.00	1.01/1.00	1.03/0.89
wave5	0.52	1.00/1.00	1.00/1.00	1.04/0.87
turb3d	0.79	1.04/0.95	1.04/0.95	1.05/0.94
fpppp	1.00	1.00/1.00	1.00/1.00	1.06/0.95

that algorithm *hsu* chose different program regions to slow down in seven of the 10 benchmarks. Running the reference data input as the training input does not necessarily yield a better result, for example, apsi. We suspect that the instrumented program for profiling has somewhat altered the instruction access pattern and is considerably different from the original code. According to Hsu's dissertation [21], the SUIF2 compiler infrastructure, on which algorithm *hsu* was built, also has a major impact on the experimental results.

5.5   A Comparison with Choi et al.'s Work

In this section, we compare and contrast our work with Choi et al.'s work in [6,7]. Recall that both works are based on the same Equation (1). The difference is in the calculation of equation coefficients. Our work calculates β defined in Equation (4), whereas Choi et al.'s work calculates α_f defined as follows:

$$\alpha_f = f \cdot \frac{T_{mem}}{W_{cpu}} \tag{8}$$

Analytically, the two metrics are equivalent:

$$\beta = \frac{1}{1 + \alpha_f \cdot f_{max}/f} \tag{9}$$

However, there are several major differences in terms of implementation. First, the β metric is invariant to a CPU frequency change, whereas the α_f metric is defined with respect to a particular CPU frequency f. Thus, the number of coefficients calculated in Choi et al.'s DVS algorithm is more than the number of coefficients calculated in algorithm β-*adaptation*. Second, the formula in calculating α_f is more complex. This is due to the two-coefficient model they use, in contrast to the one-coefficient model we use. Finally, the number of PMU event counts needed for calculating β is smaller than that for calculating α_f. Since a CPU can simultaneously count a finite number of events, counting too many events may introduce a larger time overhead.

Finally, our new DVS algorithm has a simpler implementation than Choi et al.'s work. However, we cannot do an empirical comparison given the current setting we have. Choi et al. implemented their DVS algorithms on Intel Xscale-based processors which does not provide counting for the number of *retired* instructions. On the other hand, our hardware platform, Athlon XP processor, does not provide counting for the number of *executed* instructions. In fact, this is one of the big issues in using the PMU to assist DVS scheduling — the CPU events may not be compatible nor consistent across different hardware platforms. This is also why Choi et al. presented two platform-dependent implementations [6,7] of the same DVS algorithm [6].

5.6   Sensitivity Analysis of Algorithm Parameters

In this section, we present a sensitivity analysis of the parameters in algorithm β-*adaptation*, i.e., δ for the relative performance slowdown and I for the length of an interval, as shown in Figure 1.

For the **SPEC** CPU95 benchmarks, the average execution time increases at a pace of 3% for every 5% increase in δ, whereas the average energy consumption stays around 20% after δ passes 30%. As δ increases, the algorithm slows down CPU-bound programs which have lower performance-power ratios. Hence, setting δ at a small value such as 5% is recommended.

In terms of the interval size I, the average execution time is a U-shape curve. Since setting I to a large value, such as five seconds, did not let the program run at the converged f^* for a sufficiently long time and setting I to a small value such as 10 milliseconds introduced a significant amount of time overhead, we recommend setting I at a value between 50 milliseconds to 1 second.

6   Conclusions and Future Work

In this paper, we proposed a new, PMU-assisted, interval-based, DVS algorithm that detects the CPU-boundedness of a program on the fly and adjusts the CPU speed accordingly. The algorithm is no arbitrary heuristic. It is based on an extension of the previous theoretical work for energy-optimal DVS scheduling problem. The algorithm has also proven to be effective in comparison with a number of DVS algorithms through physical measurements. That is, the new algorithm provides fine-grained, tight control over performance loss as well as exploits the sublinear performance slowdown. Finally, the algorithm is simple to implement.

Our new DVS algorithm can be refined in various ways. One particular direction is to use compiler hints as additional scheduling support. While this idea is not new (e.g., [19,22]), the type of hint that the compiler should provide so that the overall DVS algorithm is effective is still a research topic for general-purpose systems. To relieve the compiler from the difficulty of giving exact timing information off-line, we could have the compiler simply identify and distinguish execution phases of a program in terms of CPU-boundedness in an approximate manner. Algorithm β-*adaptation* can then be refined to compute the β value for each of these phases to further improve its effectiveness for memory-bound programs.

References

1. J. Lorch and A. Smith. Improving dynamic voltage algorithms with PACE. *International Conference on Measurement and Modeling of Computer Systems*, June 2001.
2. A. Miyoshi, C. Lefurgy, E. Hensbergen, and R. Rajkumar. Critical power slope: Understanding the runtime effects of frequency scaling. *International Conference on Supercomputing*, June 2002.
3. W. Kim, J. Kim, and S. Min. Preemption-aware dynamic voltage scaling in hard real-time systems. *International Symposium on Low Power Electronics and Design* , August 2004.
4. C.-H. Hsu and U. Kremer. The design, implementation, and evaluation of a compiler algorithm for CPU energy reduction. *ACM SIGPLAN Conference on Programming Languages Design and Implementation*, June 2003.

5. H. Li, C.-Y. Cher, T. Vijaykumar, and K. Roy. VSV: L2-miss-driven variable supply-voltage scaling for low power. *International Symposium on Microarchitecture*, December 2003.
6. K. Choi, R. Soma, and M. Pedram. Fine-grained dynamic voltage and frequency scaling for precise energy and performance trade-off based on the ration of off-chip access to on-chip computation time. *Design Automation and Test in Europe Conference*, February 2004.
7. K. Choi, R. Soma, and M. Pedram. Dynamic voltage and frequency scaling based on workload decomposition. *International Symposium on Low Power Electronics and Design* , August 2004.
8. F. Yao, A. Demers, and S. Shenker. A scheduling model for reduced cpu energy. *IEEE Annual Symposium on Foundations of Computer Science*, October 1995.
9. T. Ishihara and H. Yasuura. Voltage scheduling problem for dynamically variable voltage processors. *International Symposium on Low Power Electronics and Design* , August 1998.
10. K. Seth, A. Anantaraman, F. Mueller, and E. Rotenberg. FAST: Frequency-aware static timing analysis. *International Real-Time Systems Symposium* , December 2003.
11. C.-H. Hsu, U. Kremer, and M. Hsiao. Compiler-directed dynamic frequency and voltage scheduling. *Workshop on Power-Aware Computer Systems*, November 2000.
12. T. H. Cormen, C. E. Leiserson, and R. L. Rivest. *Introduction to Algorithms*. MIT Press, Cambridge, MA, 1990.
13. N. Hirofumi, N. Naoya, and T. Katsuya. WT210/WT230 digital power meters. Yokogawa Technical Report 35, 2003.
14. The Standard Performance Evaluation Corporation. http://www.spec.org.
15. Sun Fire B100x Blade Server. http://www.sun.com/servers/entry/b100x/.
16. D. Grunwald, P. Levis, K. Farkas, C. Morrey III, and M. Neufeld. Policies for dynamic clock scheduling. *Symposium on Operating System Design and Implementation*, October 2000.
17. A. Varma, B. Ganesh, M. Sen, S. Choudhary, L. Srinivasan, and B. Jacob. A control-theoretic approach to dynamic voltage scaling. *International Conference on Compilers, Architectures, and Synthesis for Embedded Systems*, October 2003.
18. N. AbouGhazaleh, D. Mossé, B. Childers, and R. Melhem. Toward the placement of power management points in real time applications. *Workshop on Compilers and Operating Systems for Low Power*, September 2001.
19. A. Azevedo, I. Issenin, R. Cornea, R. Gupta, N. Dutt, A. Veidenbaum, and A. Nicolau. Profile-based dynamic voltage scheduling using program checkpoints in the COPPER framework. *Design Automation and Test in Europe Conference*, March 2002.
20. B. Childers, H. Tang, and R. Melhem. Adapting processor supply voltage to instruction-level parallelism. *Kool Chips Workshop*, December 2000.
21. C.-H. Hsu. *Compiler-Directed Dynamic Voltage and Frequency Scaling for CPU Power and Energy Reduction*. PhD thesis, Department of Computer Science, Rutgers University, New Brunswick, New Jersey, June 2003.
22. N. AbouGhazaleh, D. Mossé, B. Childers, R. Melhem, and M. Craven. Collaborative operating system and compiler power management for real-time applications. *Real-Time Embedded Technology and Applications Symposium*, May 2003.

Appendix

To prove Theorem 1, we first show that the following chain of inequalities is true.

$$0 \geq \frac{E_2 - E_1}{T_2 - T_1} \geq \frac{E_3 - E_2}{T_3 - T_2} \geq \cdots \geq \frac{E_n - E_{n-1}}{T_n - T_{n-1}}$$

This is not difficult to prove because

$$\frac{E_i - E_{i-1}}{T_i - T_{i-1}} - \frac{E_{i+1} - E_i}{T_{i+1} - T_i} = f_i \cdot \left(\frac{P_{i+1} - P_i}{f_{i+1} - f_i} - \frac{P_i - P_{i-1}}{f_i - f_{i-1}} \right)$$

$$+ f_i \cdot \frac{c_0}{c_1} \cdot \left(\frac{P_{i+1} - P_i}{f_{i+1} - f_i} \cdot f_{i+1} - \frac{P_i - P_{i-1}}{f_i - f_{i-1}} \cdot f_{i-1} \right) \geq 0$$

and

$$\frac{E_{i+1} - E_i}{T_{i+1} - T_i} = \frac{f_i f_{i+1}}{f_i - f_{i+1}} \cdot \left[\left(\frac{P_{i+1}}{f_{i+1}} - \frac{P_i}{f_i} \right) + \frac{c_0}{c_1} (P_{i+1} - P_i) \right] \leq 0.$$

Then we define $r_i = t_i / T_i$ and introduce a new function $E_{min}(d)$ as follows.

$$E_{min}(d) = \min\{\sum_i r_i \cdot E_i : \sum_i r_i \cdot T_i = d, \sum_i r_i = 1, r_i \geq 0\}$$

Since the sequence $\{\frac{E_{i+1}-E_i}{T_{i+1}-T_i}\}_{i=1,\cdots,n-1}$ is non-increasing, function $E_{min}(d)$ is equivalent to the piecewise-linear function that connects points $\{(T_i, E_i)\}$. Since the slopes of chords in this piecewise-linear function are all non-positive, $E_{min}(d)$ is non-increasing. Thus, we seek for a solution of $E_{min}(D)$ as $E_{min}(D) \equiv \min\{E_{min}(d) : d \leq D\}$. For $T_{j+1} < D \leq T_j$, $E_{min}(D)$ is the function value at D in the chord connecting points (T_j, E_j) and (T_{j+1}, E_{j+1}). The proof is completed by solving the linear system of $t_j^* + t_{j+1}^* = D$ and $t_j^*/T_j + t_{j+1}^*/T_{j+1} = 1$.   □

Safe Overprovisioning: Using Power Limits to Increase Aggregate Throughput*

Mark E. Femal and Vincent W. Freeh

Department of Computer Science,
North Carolina State University
{mefemal, vwfreeh}@ncsu.edu

Abstract. Management of power in data centers is driven by the need to not exceed circuit capacity. The methods employed in the oversight of these power circuits are typically static and ad-hoc. New power-scalable system components allow for dynamically controlling power consumption with an accompanying effect on performance. Because the incremental performance gain from operating in a higher performance state is less than the increase in power, it is possible to *overprovision* the hardware infrastructure to increase throughput and yet still remain below an aggregate power limit. In overprovisioning, if each component operates at maximum power the limit would be exceeded with disastrous results. However, safe overprovisioning regulates power consumption locally to meet the global power budget. Host-based and network-centric models are proposed to monitor and coordinate the distribution of power with the fundamental goal of increasing throughput. This research work presents the advantages of overprovisioning and describes a general framework and an initial prototype. Initial results with a synthetic benchmark indicate throughput increases of nearly 6% from a staticly assigned, power managed environment and over 30% from an unmanaged environment.

Keywords: Overprovisioning, managing power limits.

1 Introduction

Our primary motivation is to increase throughput, given defined power limits, by increasing parrallelism. High performance clusters such as BlueGene/L [1] make use of low-power, modest clock rate processors to provide more efficient performance with respect to energy consumption. A similar approach can be taken with frequency scalable CPUs and general-purpose hardware. The CPU is a dominant power consumer in most servers and is consequently our initial focus. In addition, there is a manufacturer commitment towards CPU power conservation as exhibited in the ACPI Specification [2]. In general, scaling the the processor from higher to lower power gears tends to slow down power usage in other devices. The relationship of the CPU frequency (F) and voltage (V) is given by [3]:

$$CPU\ Power = A \cdot C \cdot V^2 \cdot F$$

* This research was supported in part by an IBM UPP award.

B. Falsafi and T.N. Vijaykumar (Eds.): PACS 2004, LNCS 3471, pp. 150–164, 2005.

Where A is the activity factor for how frequently gates switch and C is the total capacitance at the gate outputs. Frequency is proportional to voltage; therefore, CPU power is proportional to F^3. However, to the first approximation, performance is proportional to F.

Many large data centers have a goal of managing instantaneous power consumption. The physical infrastructure of data centers is typically partitioned into a set of circuits. These circuits, allocated per rack, provide a maximium quantity of instantaneous power. Data center personnel approach the problem of assigning equipment to racks (provisioning) conservatively. Exceeding the circuit power limit can cause a disruption in service. Therefore, managing instaneous power consumption is often a higher priority than reducing energy consumption.

Complicating the delicate balance of maintaining a safe upper bound on power consumption, significant variation occurs based on the state of connected equipment. A server that boots needs near its maximum rated amount of power. In contrast, lightly loaded and idle servers draw significantly less power. An analysis of work performed and power consumed facilitates safe overprovisioning. This benefit offers data centers the ability to maintain peak performance in defined power limits.

The general concept of overprovisioning is not new in industry. For instance, equipment exists to establish a sequence of power up when disruptions occur along with the ability to monitor aggregate power usage [4]. Hardware provides a means of reacting to power failures and monitoring circuit health; however, it is not a good mechanism for controlling power in a concerted fashion. Such intelligence must be used in environments that suffer from the inability to expand their power infrastructure.

We regard saving energy as a secondary goal in intelligent power allocation. In addition, we choose to not require changes to existing applications. We do not preclude possible gains by allowing applications more control in the decision process but believe solutions that require this to function are too restrictive. Our initial implementation results reflect at least a 6% gain in throughput in a synthetic benchmark while still remaining below a fixed aggregate power limit. This performance gain includes a static analysis done to ensure cluster nodes are operating at the best gear. In unmanaged environments, using the same benchmark, our implementation provides a throughput gain in excess of 30%.

In section 2 we provide an overview of the principles and design aspects of our model. Section 3 discusses the implementation. Section 4 illustrates preliminary results. Section 5 presents related work and section 6 outlines conclusions and future efforts.

2 Overview

It is useful to first formalize the motivations mentioned previously. Given a defined power limit P_{global}, a finite number of nodes can complete work subject to this constraint. If we denote P_{gear} as the power consumption of a node in a given performance gear, the number of total nodes (all nodes execute in the same gear), can be represented as $N_{gear} = \lfloor P_{global}/P_{gear} \rfloor$. Next, given the throughput for a node in a given gear represented by the function $T(P_{gear})$, total throughput of all nodes is simply $T_{total} = T(P_{gear}) \cdot N_{gear}$. The *energy efficiency* per node, *i.e.*, the useful work

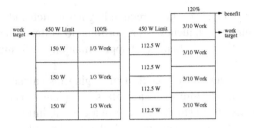

Fig. 1. Managing power to a 450 watt limit based on a 10% reduction in original work and 25% reduction in node power

per unit energy over time t is, $E_{node} = T(P_{gear})/(P_{gear} \cdot t)$. The overall efficiency of n nodes is $E_{total} = T_{total}/(n \cdot P_{gear} \cdot t)$. A known relationship between P_{gear} and $T(P_{gear})$, provides the impetus for increasing overall throughput. Modeling theoretical power usage with P_{gear} is simplified above for clarity. In practice, with varying workloads the maximum, minimum, and average power usage can differ significantly. In addition to finding N_{gear}, it is possible for a mixed set of P_{gear} values to maximize P_{global}. In general, there is a work benefit when $\Delta P_{global} > \Delta T_{total}$. If $P_{global} \, \alpha \, F^{\kappa}$, when $\kappa \leq 1$ and n increases there is a gain in E_{total}.

Figure 1 depicts the aforementioned general strategy. In a given interval, the total power available to all nodes is constrained by a 450 Watt limit. Each node on the left executes unconstrained and consumes 150 Watts while completing $1/3$ of the work. In contrast, the overprovisioned nodes on the right are now restricted to $1/4$ of the power budget so consume 25% less power per node. If the restricted nodes execute with a 10% penalty in work contribution, a reasonable expectation, these nodes provide a realized benefit of 20% due to exceeding the original work target. This example is presented for illustrative purposes and it provides the incentive to manage power consumption to increase throughput.

2.1 Local Power Limit

The locally assigned power limit for each participant in the network is regarded as a mutual decision based on global constraints. Given the global budget, each node is responsible for the suballocation of power at a fine-grained level. Each node operates within its derived maximum power limit. It is free to choose where to set its own target usage level based on need, priority, or other relevant measures.

With this allocation in mind, it is possible for the global power allocation mechanism to account for each node's power need based on the relative difference between its target usage and current limit. This allows flexible policies to be deployed at the local level but still utilizes the same simple interface for the allocation of the global budget. The global budget can be considered to be dynamically assigned based on intelligent devices on the network. For instance, smart racks or uninterruptible power supplies may supply the information for autonomic operation.

At the node's architectural level, each device in a server has an intelligent interface to relay or provide information related to system power draw, available gears, as well as minimum and maximum power requirements directly to power management algorithms. Although elements of this are available in laptop systems using the ACPI specification [2], future development of consistent interfaces to hardware should help promote similar access to server sensors using the same specification.

2.2 Design

Our general design approach to power distribution is encapsulated with the producer-consumer model. The power producers are regarded as the suppliers of power to various components. The producer at the cluster node level, such as its power supply, may in fact be a consumer at the global level but the approach is conveniently generalized in a hierarchical manner. Thus, each device in the overall system fits in this hierarchy. The policies for each device must be flexible enough to adapt to the unique demands for services from that device. The interface between devices is through its budget and aggregated at upper levels as applicable. The units of budget allocation align with measurement methodology (watts).

2.3 Components

A number of logical components comprise the framework and are shown in Figure 2. Starting at the lowest level, device controllers provide the intelligence to determine the performance gear used. Each server node has this functionality aggregated into a cohesive entity referred to as the Local Power Agent (LPA). It is responsible for determining the average power consumption target. There is loosely coupled communication between the LPA and the devices it manipulates with the device controller. A logical Power Message Bus (PMB), implemented as a message queue, facilitates communication between device drivers and the LPA. This flexibility allows for direct control by the LPA as well as the potential for negotiation of power between device drivers.

Another major component of the system is the Global Power Agent (GPA). The GPA is responsible for the coordination and interchange of related messages between nodes. It analyzes messages from the network and makes the appropriate requests to the LPA using the PMB. This assignment is possible due to all participants in a Power Management Group (PMG) broadcasting relevant information to all nodes in the cluster. There is not a one-to-one correspondence between a PMG and a subnet; however, our initial implementation limits servers in a PMG to be on the same broadcast network.

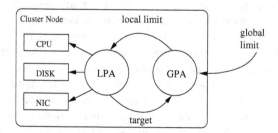

Fig. 2. Relationship of the GPA, LPA, and power-managed devices

In addition to receiving PMG broadcasts, the GPA responds to other network requests using the Power Exchange Protocol (PEP). These messages include administrative control requests (*i.e.*, setting the global power limit) in our initial prototype.

2.4   Architecture

For the initial development and model analysis, a cluster of servers was built using frequency scaling processors. Ten nodes were assembled using the following hardware: 40 GB Maxtor EIDE 7200 RPM disk drives, ASUS K8V motherboards (on-board 1Gb NIC), 1 GB of PC3200 DDR SDRAM, and an AMD64 3000+ CPU. All nodes were interconnected on a dedicated 100 Mb switch. Although we regard power distribution as a general allocation problem across a diverse set of hardware architectures and computational resources, our initial implementation is limited due to budgetary reasons.

The entire cluster uses the Linux 2.6 kernel. For frequency and voltage scaling, the AMD PowerNow *cpufreq* module is used. Modifications were done to augment the ACPI device tables in the BIOS. These additional performance gears were subsequently used in the *cpufreq* driver as indicated in Table 1. Measured system idle power consumption is shown along with expected CPU power usage obtained from [5]. System idle power reflects reduced power usage from using the HLT instruction. The modified frequency and voltage settings were used to expose additional gears for the purpose of our evaluation.

Table 1. AMD64 3000+ CPU and system idle power consumption

Frequency (Mhz)	BIOS Voltage	CPU (Watts)	New Voltage	System Idle (Watts)
2000	1.5	89	1.5	89
1800	1.4	66	1.4	86
1600	-	-	1.35	84
1400	-	-	1.3	83
1200	-	-	1.2	81
1000	1.1	22	1.1	79
800	-	-	1.0	77

For calculating system power measurements, two digital multimeters (DMMs) were connected to serial ports on a non-cluster server. Custom software was created to interface with the DMMs located on this host. One meter was configured to measure AC voltage and the other inserted in a serial fashion to measure amperes on a single node. To allow a cluster node to have a notion of power usage, a TCP-based request server was created to allow a node to query power usage as needed. The overhead of this network access is relatively small and a mechanism such as this is needed due to the lack of hardware sensors on each node. In the future, we plan to augment this framework using a more robust collection mechanism for independent cluster node measurement. Our initial research focus is currently on intra-node decisions so the power measurement method is not a limiting factor. Single node measurements are used to extrapolate expected system power usage for multiple nodes.

3 Implementation

As previously discussed in Section 2.3, there are two primary components in our implementation for managing power. The GPA assigns the local power limit based on the aggregate circuit capacity and information received from all cluster nodes. The LPA is responsible for ensuring the target power usage of an individual node is met. Each of these is a separate daemon process and both are implemented as non-priviledged processes.

3.1 Local Power Agent

The LPA is the aggregated power control framework for all devices in a cluster node. It is responsible for maintaining a power target and listening for inbound communication on the PMB. Each device has its inherent power characteristics coordinated with other devices by this agent. The PMB provides an abstraction between the idiosyncrasies inherent in each device and the general controller routine. Each device has several gears the LPA manipulates to control overall system power using a device specific routine exposed to the controller.

To simplify the interface to device gears, we regard the lowest power usage gear as the highest numbered performance gear. Thus, performance gear 0 in the state array for each device is the highest power usage gear. Although this is our internal convention, we still refer to increasing the performance gear as an increase in the power usage gear.

The primary focus of the LPA is determining the target power based on the limit assigned by the GPA. This derivation remains flexible to have different policies implemented depending on the behavior desired. Two sample policies include one based on overall load and another might be to optimize for a performance delay characteristic. This policy is not restricted to a single rule, a combination of rules could certainly be employed.

A feedback controller manipulates device gears to meet the target system power set by the policy. To maintain the local power limit, the controller employs a predictor to determine the expected usage in the next epoch. We regard the power limit as an upper bound on instantaneous power usage. A sampling window of size w facilitates keeping system power close to the target. In the initial implementation, the window size is 30 seconds. To prevent excessive gear switching and allow stabilization, a minimum time between changes is enforced. This delay also helps manage the differing capabilities of devices and their subsequent ability to transition to different gears in a specified time interval.

The core controller uses a PID algorithm [6]. If system power is denoted as S and the average power target is μ, the error is $\epsilon = \mu - S$. Next, the controller calculates joules used in the epoch using $j_e = t_e \cdot \epsilon$. With instantaneous error known and gain constants G_1, G_2, and G_3, the energy surplus or deficit is

$$\eta_e = G_1 \cdot j_e + G_2 \cdot \int_0^w j_e dt + G_3 \cdot (j_e - j_{e-1}).$$

With η_e known, a prediction for the next epoch is determined to ensure the node power limit is not exceeded. For this prediction δ, we find $\delta = max(j_e - j_{e-1}, \delta)$.

A non-uniform distribution of δ values exist for each performance gear. Thus, a table is managed at runtime that represents the quantization of discrete power step values at a resolution defined at compile time for each gear. It is pessimistic due to using the highest value seen in subsequent predictions. The estimator P_e for the expected power usage in the next epoch is then found using

$$P_e = S + \frac{\delta + ((j_e - j_{e-1}) - (j_{e-1} - j_{e-2}))}{t_e}.$$

The LPA does not have a strict power limit. Instead, a burstable region exists above and below the GPA assigned limit. The impact to the feedback controller is that if given a burst allocation of τ and limit of ω, it is $\tau + \omega$ that is regarded as the true limit. The burst allocation allows nodes to respond more rapidly to workload demands in lower performance gears. The current implementation regards the burstable region above the power limit as a soft limit so P_e is compared to $\tau + \omega$ to determine if an immediate gear reduction is needed. If no immediate correction is necessary, η_e is checked against a threshold preset on LPA startup. If the threshold is exceeded and $\eta_e < 0$ (implies overusage) the CPU gear is decreased. The inverse condition of $\eta_e > 0$ increases the gear

3.2 Global Power Agent

The GPA's responsibility is to allocate the node power limit based on state information received from all nodes in the PMG. For reliability and scalability, each individual node is responsible for determining the power limit. Although there is explicit trust in ensuring all nodes are well-behaved, this precondition should be acceptable in many environments. All nodes in the PMG are synchronized by periodic UDP broadcasts. Nodes are added or removed from the subnet with corresponding changes done to power limits per the specific policy engine implementation.

Power Management Groups. The cluster management policy is shared on all nodes in the current implementation. To allow for multiple logical assignments and allocations on the same broadcast subnet, a cluster identifier is configured for each GPA on daemon startup and is referred to as the PMG. The cluster data structure is an AVL tree, so tree operations are bounded by $O(lg\ n)$ where n is the number of PMG nodes (non-member broadcasts are simply ignored). A dedicated thread is responsible for receiving UDP packets describing the state of other nodes in the PMG as well as reacting to administrative commands (further explained below). This thread uses *select()* with a timeout to prune the tree based on the time stamp of the last broadcast received for a cluster node and a predetermined maximum broadcast retention value. Within the current implementation, this value is 30 seconds and the broadcast rate is once per second.

Broadcast Messages. There are two types of broadcast messages currently sent to participants in a PMG. First, broadcast data packets are sent containing a node's current power usage, limit, burst, and target values. In addition, a sequence number is sent that all cluster members use to ensure decisions are computed in a coordinated fashion. This sequence number is initially generated by the first node on a subnet.

In addition to data packets from other members in the PMG, administrative messages can be broadcast to all nodes. Such notifications consist of modifications to the overall power limit to be used by the policy engine of the GPA as well as for setting an immediate administrative node limit used by a support tool. Membership in a PMG is further refined to be either *active* or *passive*. In passive mode, broadcasts are sent and received as normal, but inbound administrative messages are ignored. In active mode, the node responds to administrative messages.

3.3 Support Tools

Although the long-term intent of the power management framework is to be totally autonomic in operation, the current implementation receives instructions from support tools. Tied into the same communicative protocol as the GPA, a console application monitors the state of either a specific PMG or all nodes running on a given subnet. For administrative control of a given node, a tool exists to interface directly with the LPA (as does the GPA) through shared memory or by sending messages on the PMB. For remote requests, the tool communicates indirectly through the remote node's GPA.

4   Results

We evaluated the initial prototype and model on the architecture previously mentioned in Section 2.4. To examine the nature of the tradeoff between throughput and power, we first constructed a series of synthetic CGI [7] programs runnable by an Apache 2.0 web server. We used *httperf* [8] to generate a sustained workload by using four cluster nodes configured as request clients. A single server was setup to handle requests and clients were configured to overload the server based on a request timeout of five seconds. When the aggregate number of errors for all clients exceeded five percent, we considered the throughput for the server to be maximized for a given performance gear. For the time alloted for a benchmark (30 seconds), the average system power consumption was calculated using the multimeters. Due to the high number of nodes needed to generate the workload effectively and current limitations measuring system power, extrapolations based on single node measurements are used to show how effective the solution performs when increasing the number of nodes. An additional cluster node is also used to control the request clients and collect the measured values for reporting purposes.

Our results notation depicts the highest performance gear (and highest power consumption gear) as zero. A consistent theme emerges from the synthesized benchmarks. The highest performing gear does not have the highest performance (*i.e.*, throughput) per unit power. To illustrate this difference in power with respect to throughput, consider Figure 3. It shows the resultant increase in power from gear 1 to 0 is approximately 12.57%; however, the throughput gain is only 6.25%. In contrast, the increase in throughput from gear 6 to 5 is 25% with only a 6.19% increase in average power usage.

Table 2 depicts the raw data collected in this benchmark. In an environment such as this, it is possible to reduce the CPU performance gear to conserve power and increase the number of nodes to service that load and still decrease total power consumption. This is shown by a simple example and the data in Table 2. If the service

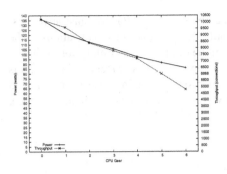

Fig. 3. Merge sort power vs. performance trade-off

Table 2. Power and throughput details for merge sort

Gear	Throughput	Change	Power	Change
0	10200	6.25%	136.63	12.57%
1	9600	14.29%	121.37	7.65%
2	8400	7.69%	112.74	6.41%
3	7800	8.33%	105.95	8.49%
4	7200	20.00%	97.66	6.66%
5	6000	25.00%	91.56	6.19%
6	4800		86.22	

Table 3. Throughput gains with a 600 watt power limit and static gear assignment

Nodes	Gear	Throughput	Power	Gain
4	0	40800	546.52	
4	1	38400	485.48	-0.06%
5	2	42000	563.70	0.03%
6	3	46800	529.75	12.82%
6	4	43200	585.96	5.56%
6	5	36000	549.36	-11.76%
6	6	28800	517.32	-41.67%

Table 4. Power and throughput details for insertion sort

Gear	Throughput	Change	Power	Change
0	10800	12.5%	140.87	11.40%
1	9600	6.67%	126.45	6.87%
2	9000	7.14%	118.32	6.06%
3	8400	7.69%	111.56	8.23%
4	7800	18.18%	103.08	7.84%
5	6600	37.50%	95.59	6.91%
6	4800		89.41	

requirement states the desired throughput is 36000 connections in 30 seconds (with a client request timeout of 5 seconds), 5 cluster nodes as configured in our architecture meet this requirement running in gear 4. The subsequent total power usage is approximately $5 \cdot 97.66 = 488$ watts. Four similarly configured servers (without managing the power proactively) could also service this same load; however, the total power usage is

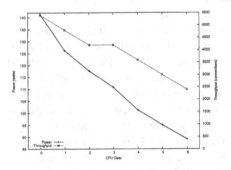

Fig. 4. Prime numbers under 10000 power vs. performance trade-off

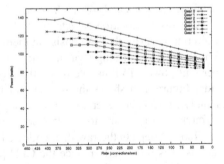

Fig. 5. Relationship of power and concurrency to print environment variables

Table 5. Fixed gear throughput, aggregate power, and ideal node power usage ($P_{gear} = P_{global}/N$).

N	Gear	Throughput	Power (W)	P_{gear} (W)
3	0	62745	425.43	183.33
4	1	77184	507.92	137.50
4	2	71728	476.96	137.50
4	3	64052	444.96	137.50
5	4	71895	514.40	110.00
5	5	57805	480.10	110.00
6	6	53652	541.38	91.67
10	0	209150	1418.10	150.00
11	1	212256	1396.78	136.36
12	2	215184	1464.32	125.00
13	3	208169	1446.12	115.38
14	4	201306	1440.32	107.14
15	5	173415	1440.30	100.00
16	6	143072	1443.68	93.75

Table 6. Dynamic gear throughput and aggregate power using the LPA ($\omega = P_{gear}$).

N	Throughput	Power (W)	ω (W)
4	81600	546.96	137.50
5	81745	547.70	110.00
6	63204	544.86	91.67
11	224103	1498.09	136.36
12	229584	1498.68	125.00
13	216112	1464.32	115.38
14	210826	1477.84	107.14
15	182700	1500.00	100.00
16	141232	1478.88	93.75

$4 \cdot 136.63 = 547$ watts. This example illustrates a decrease of about 11% in total power needed. Of the 10 benchmarks created, 9 exhibit the general trend depicted in Figure 3.

The preceding example used an increase in nodes to show the benefit in conforming to a lower power limit. In addition, this same data is used in Table 3 to reflect additional gains in throughput. Notice the gain in throughput at gear 3 is nearly 13% despite the 70.25 Watts of unused power. Without a more proactive means of power allocation (*i.e.*, an LPA) this represents a loss in throughput.

The preceeding examples help motivate our research but do not reflect the strict representation of all applications. For instance, an exception to the high difference of power and throughput at gear 0 was exhibited in one of our synthentic benchmarks. The resulting power-throughput relationship in Table 4 shows an even larger gain in throughput at lower power gears (*i.e.*, 6 to 5) than the preceeding example. On the other hand, the benefit of the reduced power gear 1 (from 0) is not as pronounced.[1]

Another important result emerged from the affects of one benchmark. Figure 4 reflects no gain in throughput going from gear 3 to 2. Unfortunately, there is a 6.21% increase in power usage for this transition. The data reveals the lower bound overload threshold was just exceeded for gear 3 and the upper bound prevented gear 2 from additional gains. With this result, a power management policy should select the lower power gear.

Rather than just considering total throughput in a given time interval, one can also analyze the power needed to sustain a desirable concurrency. Figure 5 shows an example benchmark illustrating power needed in all seven gears. Depicting the highest performance on the left, in the slowest gear there is a 27.3% increase in concurrency to go to the next highest gear with a corresponding increase of 6.5% in power usage. This

---

[1] Sort-based benchmarks used the same quantity and distribution of items in all algorithms.

is noted by comparing the leftmost points of the bottom two curves. At higher performance gears the additional gains in concurrency are perhaps not worth the additional power consumption. The concurrency difference is only 5% transitioning from gear 1 to 0 with an increase of 10.9% in power consumption. An opportunity cost analysis such as this can be used in a policy controller that attempts to meet a given level of concurrency.

To verify the efficacy of the LPA, the merge sort CGI is used as a representative benchmark. In the first data set, a 1500 Watt limit is established and extrapolation starts with 10 nodes. This allows an interpolated point to occur between each static gear assignment. In the second data set, a 550 Watt limit is used and extrapolation starts at 3 nodes. To reduce variability and ensure an adequate length of time for the test, throughput measurements were obtained after sustained load was applied for 60 seconds. Throughput was again considered maximized based on a 5 second timeout using 4 cluster nodes as request clients. Table 5 depicts the results of the static assignment. Total effective throughput is found by subtracting the number of errors from the total connections. Based on the number of nodes and the ideal power limit, the interpolated results using the LPA are shown in Table 6.

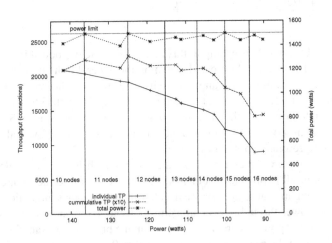

Fig. 6. LPA performance with target power set to P_{gear} using 1500 Watts starting with 10 nodes

Figures 6 and 7 illustrate the benefit of overprovisioning using the data from Tables 5 and 6. There are two lines depicting throughput, one for an individual node and the other for extrapolated aggregate values. As expected, the individual throughput decreases as available power is reduced. The aggregate throughput curve shows the performance of a cluster. When the power decrease crosses a vertical bar representing the ideal power usage, another node is supported within the power limit. It is important to note that the highest cluster throughput is not the best performance state in both figures.

As the raw data in Tables 5 and 6 show, there is a 7.15% gain in throughput between gears 0 and 1 for 11 nodes. In addition, there is a 6.69% gain in throughput between a staticly assigned gear 2 and the LPA using 12 nodes. Notice that after gear 2, there

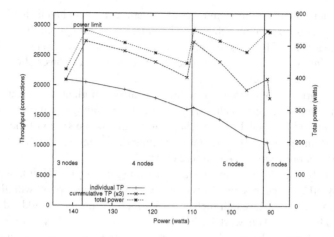

Fig. 7. LPA performance with target power set to P_{gear} using 550 Watts starting with 3 nodes

is no resultant benefit from increasing the number of nodes. Additional consideration is warranted if considering an environment, subject to this same limit, that does no proactive management of performance gears. In such a case, only 10 nodes can exist under the limit so the resulting throughput benefit of the LPA using 12 nodes is 9.77%. Clearly these benefits are dependent on the power capacity available and the number of nodes. If starting with 3 nodes, the benefit from a statically assigned solution is approximately 5.91% using 4 nodes at gear 1; moreover, the gain from an unmanaged 3 node solution is 30.28% with a cluster of 5 LPA managed nodes.

5   Related Work

The case for a closer relationship between the operating system and power management is explored in [9,10]. Flinn and Satyanarayanan [11,12] show that coordination with applications can yield significant power savings. Dynamic voltage scaling (changing both frequency and voltage) to reduce power consumption was explored in [13, 17] A vast amount of research has been done with regards to energy conservation (EC) on mobile platforms [18, 22]. The role of EC is complimentary to our goal of efficient power allocation. Power represents the instantaneous energy used at a specific point in time and energy is the usage of power in a defined interval. Our strategy maximizes throughput constrained by the limit using a target power. This power target could be derived from an EC policy; however, it is not a necessary condition.

Power management in commercial servers is important for web servers [23,24]. Much of this work relies on load balancers to distribute work. An investigation of load balancing was done in [25,26] to turn cluster nodes on or off based on load. Additional research has also been done by Elnozahy *et al.* [27] for developing mechanisms for energy-efficient clusters using combinations of IVS, CVS, and VOVO policies. Although the VOVO policy is not considered in our initial implementation, its importance is less significant in a mixed architecture environment. In [28], an economic approach is chosen to determine the minimal number of servers required to handle load. Unlike [28] we favor a decentralized approach and seek to maximize throughput. In [29], Sharma *et al.* applies real-time techniques to web servers in order to conserve energy and maintain QoS. Managing to service metrics is one instance of a target power allocation mechanism in our local controller.

In server farms, disk energy consumption is also important. One study of four energy conservation schemes concludes by stating that reducing spindle speed is the only option for clusters [30]. DRPM is a scheme to modulate the speed of the disk dynamically to save energy [31,32] rather than stopping disk rotation. We plan to investigate this approach in our framework in future efforts.

While analyzing energy efficiency and operating points in [33], it is found that the most energy efficient gear is not always the lower performance point with their concept of *critical power slope*. Research such as in [34] focus on uniform workload distribution with migratable loads. This work has a contribution towards policy development in the management of power limits. The approach of estimating power consumption using performance counters is taken in [35, 37] and is complimentary to our notion of target power assignment.

6   Future Work

This paper has investigated the potential of overprovisioning in data centers with a focus on increasing throughput based on a defined power limit. We have presented the preliminary framework used for our research but it only represents an initial foray into the realm of distributed and local power control. Consistent with our approach to make the overall local and global mechanisms perform in a "hands free" operation, we are investigating adjustments to our local controller that will tune it automatically based on the demands of individual devices. Additional work is also needed with regards to optimizing local and global target power policies. Although target power usage has been explored, research on regarding the power limit as an unsurpassable barrier is underway along with providing intra-node differentiated service.

The status of our GPA is constrained due to a limited system for power measurement. A robust implementation is planned to enable future emphasis on nonuniform workloads and heterogeneous architectures. Greater focus and future efforts will also address current shortfalls with regards to initial node power on, synchronization of broadcast data and timing irregularities, and simultaneous cluster node adjustments. The latter will enable us to maintain a tighter upper bound on the true global power limit. The Power Exchange Protocol will also be enhanced with additional capabilities to request or deliver power to any node in the cluster within a transactional framework. This is planned for the direct reallocation of power to handle predictive power conditions not accounted for in the global allocation strategy.

We have presented a framework to handle safe overprovisioning based on power limits. We have motivated the objective and goals for our research based on preliminary data collected. The additional throughput gains possible from a strategy to manage the power limit can help increase the computational effectiveness of data centers without causing difficulties with existing power infrastructure.

References

1. Adiga et al., N.: An overview of the BlueGene/L supercomputer. In: Supercomputing 2002. (2002)
2. http://www.acpi.info: Advanced Configuration and Power Interface Specification, Revision 3.0. Hewlett-Packard Corporation, Intel Corporation, Microsoft Corporation, Phoenix Technologies Ltd., and Toshiba Corporation (2004)
3. Mudge, T.: Power: A first class architectural design constraint. IEEE Computer **34** (2001) 52–57
4. American Power Conversion Corp.: APC power distribution products. `http://www.apc./products/` (2004)
5. Advanced Micro Devices, Inc.: AMD Athlon 64 processor data sheet. `http://www.amd.com/us-en/assets/content_type/white_papers_and_tech_docs/24659.24659.PDF` (2004)
6. Minerick, R.J., Freeh, V.W., Kogge, P.M.: Dynamic power management using feedback. In: Workshop on Compilers and Operating Systems for Low Power, Charlottesville, Va (2002) 6–1–6–10
7. University of Illinois at Urbana Champaign: The common gateway interface. `http://hoohoo.ncsa.uiuc.edu/cgi/` (2004)
8. Mosberger, D., Jin, T.: httperf: A tool for measuring web server performance. In: WISP, Madison, WI (1998) 59–67
9. Vahdat, A., Lebeck, A., Ellis, C.: Every joule is precious: The case for revisiting operating system design for energy efficiency. In: Proceedings of the 9th workshop on ACM SIGOPS European workshop. (2000) 31–36
10. Ellis, C.: The case for higher-level power management. Proceedings of the 7th Workshop on Hot Topics in Operating Systems (1999)
11. Flinn, J., Satyanarayanan, M.: Energy-aware adaptation for mobile applications. In: Symposium on Operating Systems Principles. (1999) 48–63
12. Flinn, J., Satyanarayanan, M.: Powerscope: A tool for profiling the energy usage of mobile applications. In: Proceedings of the Second IEEE Workshop on Mobile Computing Systems and Applications. (1999)
13. Flautner, K., Reinhardt, S., Mudge, T.: Automatic performance-setting for dynamic voltage scaling. In: Proceedings of the 7th Conference on Mobile Computing and Networking MOBICOM '01. (2001)
14. Gruian, F.: Hard real-time scheduling for low-energy using stochastic data and DVS processors. In: Proceedings of the International Symposium on Low-Power Electronics and Design ISPLED '01. (2001)
15. Pering, T., Burd, T., Brodersen, R.: The simulation and evaluation of dynamic voltage scaling algorithms. In: ISLPED 1998. (1998)
16. Pouwelse, J., LangenDoen, K., Sips, H.: Energy priority scheduling for variable voltage processors. In: Proceedings of the International Symposium on Low-Power Electronics and Design ISPLED '01. (2001)
17. Im, C., Kim, H., Ha, S.: Dynamic voltage scheduling technique for low-power multimedia applications using buffers. In: Proceedings of the International Symposium on Low-Power Electronics and Design ISPLED '01. (2001)
18. Zeng, H., Ellis, C.S., Lebeck, A.R., Vahdat, A.: Currentcy: Unifying policies for resource management. In: USENIX 2003 Annual Technical Conference. (2003)
19. Anand, M., Nightingale, E., Flinn, J.: Self-tuning wireless network power management. In: Mobicom. (2003)

20. Helmbold, D.P., Long, D.D.E., Sherrod, B.: A dynamic disk spin-down technique for mobile computing. In: Mobile Computing and Networking. (1996) 130–142
21. Douglis, F., Krishnan, P., Bershad, B.: Adaptive disk spin-down policies for mobile computers. In: Proc. 2nd USENIX Symp. on Mobile and Location-Independent Computing. (1995)
22. Li, K., Kumpf, R., Horton, P., Anderson, T.E.: A quantitative analysis of disk drive power management in portable computers. In: USENIX Winter. (1994) 279–291
23. Bohrer, P., Elnozahy, E., Keller, T., Kistler, M., Lefurgy, C., McDowell, C., Rajamony, R.: The case of power management in web servers. In Graybill, R., Melham, R., eds.: Power Aware Computing. Kluwer/Plenum (2002)
24. Lefurgy, C., Rajamani, K., Rawson, F., Felter, W., Kistler, M., Keller, T.W.: Energy management for commerical servers. IEEE Computer (2003) 39–48
25. Pinheiro, E., Bianchini, R., Carrera, E., Heath, T.: Load balancing and unbalancing for power and performance in cluster-based systems. In: Proceedings of the Workshop on Compilers and Operating Systems. (2001)
26. Pinheiro, E., Bianchini, R., Carrera, E.V., Heath, T.: Dynamic cluster reconfiguration for power and performance. In: Compilers and Operating Systems for Low Power. (2001)
27. Elnozahy, E.M., Kistler, M., Rajamony, R.: Energy-efficient server clusters. In: Workshop on Mobile Computing Systems and Applications. (2002)
28. Chase, J.S., Anderson, D.C., Thakar, P.N., Vahdat, A., Doyle, R.P.: Managing energy and server resources in hosting centers. In: Symposium on Operating Systems Principles. (2001) 103–116
29. Sharma, V., Thomas, A., Abdelzaher, T., Skadron, K.: Power-aware QoS management in web servers. In: 24th Annual IEEE Real-Time Systems Symposium, Cancun, Mexico (2003)
30. Carrera, E.V., Pinheiro, E., Bianchini, R.: Conserving disk energy in network servers. In: Proceedings of International Conference on Supercomputing, San Fransisco, CA (2003) 86–97
31. Gurumurthi, S., Sivasubramaniam, A., Kandemir, M., Franke, H.: Dynamic speed control for power management in server class disks. In: Proceedings of International Symposium on Computer Architecture. (2003) 169–179
32. Gurumurthi, S., Sivasubramaniam, A., Kandemir, M., Franke, H.: Reducing disk power consumption in servers with DRPM. IEEE Computer (2003) 41–48
33. Miyoshi, A., Lefurgy, C., Hensbergen, E.V., Rajamony, R., Rajkumar, R.: Critical power slope: Understanding the runtime effects of frequency scaling. In: Proceedings of the 16th International Conference on Supercomputing. (2002) 35–44
34. Bradley, D., Harper, R., Hunter, S.: Workload-based power management for parallel computer systems. IBM Journal of Research and Development 47 (2003) 703–718
35. Bellosa, F.: The benefits of event-driven energy accounting in power-sensitive systems. In: Proceedings of the 9th ACM SIGOPS European Workshop. (2000)
36. Joseph, R., Martonosi, M.: Run-time power estimation in high performance microprocessors. In: Proceedings of the International Symposium on Low-Power Electronics and Design ISPLED '01. (2001)
37. Gniady, C., Hu, Y.C., Lu, Y.H.: Program counter based techniques for dynamic power management. In: Proceedings of the 10th International Symposium on High-Performance Computer Architecture. (2004)

Power Consumption Breakdown on a Modern Laptop

Aqeel Mahesri and Vibhore Vardhan

University of Illinois, Urbana-Champaign, Urbana IL 61801, USA

Abstract. The purpose of this work was to obtain a component-wise breakdown of the power consumption a modern laptop. We measured the power usage of the key components in an IBM ThinkPad R40 laptop using an Agilent Oscilloscope and current probes. We obtained the power consumption for the CPU, optical drive, hard disk, display, graphics card, memory, and wireless card subsystems–either through direct measurement or subtractive measurement and calculation. Moreover, we measured the power consumption of each component for a variety of workloads. We found that total system power consumption varies a lot (8 W to 30 W) depending on the workload, and moreover that the distribution of power consumption among the components varies even more widely. We also found that though power saving techniques such as DVS can reduce CPU power considerably, the total system power is still dominated by CPU power in the case of CPU intensive workloads. The display is the other main source of power consumption in a laptop; it dominates when the CPU is idle. We also found that reducing the backlight brightness can reduce the system power significantly, more than any other display power saving techniques. Finally, we observed OS differences in the power consumption.

1   Introduction

Mobile systems have become increasingly more powerful, but they depend on a battery, which can only power it for a limited time. To extend the battery life, we need to reduce system power without compromising performance. This has motivated newer portable computers to feature components that support several power modes. Examples include processor Dynamic Voltage Scaling (DVS), low power modes in RAMBUS DRAM, wireless card radio power modes, and others. Moreover, there is a big research initiative to exploit these component level power management features for reducing power consumption. For instance, the GRACE-OS scheduler sets the CPU speed based on application demand [1], power aware page allocation puts active pages on a minimal set of memory chips [2], and co-operative I/O queues hard disk accesses to maximize the standby time [3].

A component-wise power consumption breakdown is necessary for evaluating the actual effectiveness of these power management techniques. In some cases, component level power management techniques can potentially lead to increase in system power consumption. For example, a component that uses small fraction

B. Falsafi and T.N. Vijaykumar (Eds.): PACS 2004, LNCS 3471, pp. 165–180, 2005.

of total system power may be managed in a way that increases energy usage by other components. Moreover, a breakdown of power consumption is essential for guiding future research in power management. Researchers will want to target those components that are currently using the most power.

Our work sought to obtain this breakdown of power consumption. We obtained the power usage of the CPU, optical drive, hard disk, display, graphics card, memory, and wireless card subsystems. We further compiled this breakdown for each of a variety of workloads, to reflect how such numbers would differ for laptops being used in different environments.

To obtain the power consumption breakdown, we performed measurements in several phases. The first phase involved stripping the system to the minimum configuration that would still be usable. The next was measuring, either directly or subtractively, the power consumption of each component in all possible power modes. The third step was to run several benchmarks, and determine the component-wise power consumption. The fourth step was to determine the power consumption of the components not in the stripped system. The final step was to determine the component-wise power consumption for the workloads that used these additional components.

The main results of this study are:

- Total system power varies considerably depending on workloads.
- CPU power dominates, in spite of DVS, for many applications.
- Display power, which is affected most by backlight brightness, dominates when system is idle.
- Graphics, wireless and optical drives are major power consumers only in specific workloads.

The rest of the paper is organized as follows: Section 2 describes the experimental setup. Section 3 presents the methodology in detail. Section 4 compiles the results of our experiments and Section 5 discusses them. Section 6 examines related work and we conclude in Section 7.

2   Experimental Setup

The experimental setup consisted of three main elements, the testing platform (a laptop), the measurement apparatus, and the software to be run while doing the measurement.

2.1   Testing Platform

The measurements were performed on an IBM ThinkPad R40 laptop. The laptop was ideal for this study as it is representative of the current breed of laptops in performance and battery life (nearly five hours on a single battery). Table 1 shows the major features of this laptop.

Table 1. Test platform details

Component	Details
Processor	1.3 GHz Pentium M
Memory	256 MB
Hard Drive	40 GB @ 4200 RPM
Optical Drive	CD-R/RW, DVD
Wireless Networking	Intel Pro Wireless 2100
Screen	14.1" 1048 x 768

2.2 Measurement Apparatus

The main element of the measurement apparatus was a 50 MHz, Agilent 54621A analog oscilloscope. This oscilloscope had some very handy features such as MegaZoom, and math functions. To measure the voltage, we used a standard voltage probe. To measure current we used the Agilent N2774A current probe. This probe allowed us to measure current without breaking circuits. A limitation of this probe is that it needs to be clamped around the wire for which it is measuring current. As a result, we could not directly measure current used by devices that are directly soldered on, or are slotted into the motherboard. For most of the measurements presented, we averaged our measurements over five seconds and repeated multiple times.

2.3 Software

Since major goal of this project was to examine the workload-dependent nature of a laptop's power consumption, we used a variety of workloads to represent a wide range of tasks that may be performed on the machine. Moreover we needed several measurements for obtaining a component-wise breakdown of power consumption, and thus each workload had to be completely reproducible. Our workloads included the PCMark and 3DMark benchmarks, as well as multimedia playback and an FTP download and upload.

PCMark2002 consists of separate CPU, memory, and hard drive performance tests, as well as a combined "crunch test." These tests, while synthetic, are representative of the CPU, memory, and hard drive intensive tasks performed by typical home and office applications.

A stress test for any machine is 3D games. To measure the power consumption that might be typical during gameplay, we used the 3DMark2001 SE benchmark. This benchmark tests the performance of the CPU, memory system, and graphics controller by rendering 3D scenes representative of modern 3D games.

For measuring the power consumption that could be expected during Internet usage, we used FTP over a wireless LAN. FTP stresses the wireless LAN card and the hard drive, as is typical during Internet downloads.

For measuring the power consumed by multimedia applications, we played an audio CD. This stresses the optical drive, as well as some I/O subsystems.

3   Methodology

We initially envisioned performing direct measurements for power consumption of all components, but the limitations of the setup made this difficult. So, we divided the components into categories by the method by which they were measured:

1. Directly measurable - Hard drive, LCD/backlight, Speakers, Cooling Fan.
2. Indirectly measurable
   (a) Non-removable - Processor, Memory, Graphics.
   (b) Removable - Wireless card, Optical Drive, Modem, USB/1394 Ports (turned off)

The power consumption for the components in the first category, as well as the system power, was obtained using direct measurement with the current probe. For the second category, we used subtractive measurement. The basic idea is that for each benchmark we obtained the power consumption of the entire system with a given component in several different modes. The difference between these measurements gives the power consumption of the component.

The power measurement was conducted in several phases. The first was to strip the system to the minimum configuration that would still be usable, i.e. containing only the directly measurable and non-removable components. The next was to measure either directly or subtractively, the power consumption of these components in each possible state. The third step was to run the PCMark and 3DMark benchmarks, determining the component-wise power consumption. The fourth step was to determine the power consumption of the components not in the stripped system, including the DVD drive and the wireless LAN card. The final step was to determine the component-wise power consumption for the workloads that used these additional components.

The first phase of the power measurement was to strip the system to its minimal configuration and gaining access to the directly measurable components, which required disassembling the laptop. This disassembly was done following the steps in the notebook service manual [9]. This process was time-consuming, as we had to keep track of all the disassembled parts, and avoid static discharge once the outer case was removed. The modular design of modern notebooks has made them easier to open, but it has also them difficult to use for power measurement, as most of the modules are connected without wires (they plug into sockets that are soldered on to the motherboard).

Once the laptop was disassembled, the power supplies for the directly measurable components had to be identified and isolated. One of these components was the LCD display system. The LCD assembly on the R40 consists of two major components: the LCD panel and the inverter card that powers the backlight. The connector between the motherboard and the LCD assembly consisted of nearly 40 wires bundled together. This together with the unavailability of datasheets for the LCD panel and the backlight made the task of identifying the power supply wires challenging. However, the wires coming out of the sockets

Table 2. HDD power consumption

Hard Drive State	Power Consumption
Idle	.575 W
Standby	.173 W
Read	2.78 W
Write	2.19 W
Copy	2.29 W

on the inverter card and LCD panel were laid out separately in a transparent ribbon cable behind the LCD panel, and so after unbundling the wires we could separate them into smaller bundles by whether they went to the inverter for the backlight or to the LCD panel, and further separate these wires by the voltages we measured at their ends.

Another measurable component that did not have a separate power connector was the hard drive. In fact the hard drive was connected directly to a female connector on the motherboard, with 4 of the pins on the connector used for the power supply. To isolate these, we made an intermediate connector from two series-connected 44-pin laptop IDE cables, with the four power supply wires separated from the main cable.

We made an additional intermediate connector for the system power supply; this intermediate connector exposed the terminals of the connector to enable voltage measurement and separated the plus and minus wires to allow easy connection to the current probe.

Once we had access to the measurable component power supplies, we proceeded to measuring the power consumption of the components in the stripped system, specifically the hard drive, display, CPU, memory, and graphics controller, as well as the power loss through the system power supply.

The hard drive power was measured using the direct measurement method. The HDD on the R40 laptop had three power states, namely standby, idle and active. The standby state is the lowest power state in which the HDD motor is not spinning. Next is the idle state in which the HDD is spinning but no data is being transferred. Finally, the HDD is in the active state when data is being read or written. The power options in Windows OS allows setting the time after which the hard drive should be put to standby power mode. This value was set to three minutes, and the current going into the HDD was measured over a period of 500 seconds. The measurements showed the power state transitions, and the values obtained are summarized in the top rows of Table 2. To measure the active state power, the read, write and copy tests of the PCMark2002 benchmark suite were used. The power consumption averaged over a fixed number of reads, writes and copies is summarized in the bottom rows of Table 2.

The LCD panel operating voltage and current were directly measured using the probes. The initial measurements were done with the typical Windows background. The LCD panel current showed no variation while the backlight brightness was changed. Measurements were also obtained when LCD panel dis-

Table 3. Effect of display color on LCD power

Background Color	Power Consumption
Black	1.01 W
White	0.93 W
Windows default background	0.99 W

played a completely white, and a completely black background. Finally, color bit-depth for the pixel was varied from 16 bit to 32 bit in Windows, and from 8 bit to 16 bit to 32 bit in Linux for all the three backgrounds. The results obtained from the measurements are summarized in Table 3.

No measurable difference was seen either at the LCD panel, backlight or the system power consumption when the color bit-depth was changed.

The power consumption of the LCD backlight was obtained by performing a direct measurement on the current used by the backlight. The measurement was repeated for each of the eight available brightness levels, and the values

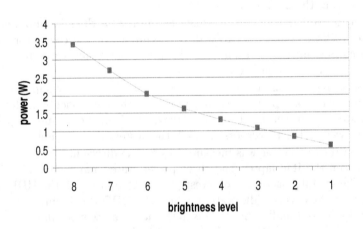

Fig. 1. Backlight power versus brightness level

Table 4. System power consumption (Watts) under Linux and Windows with CPU idle and maximized

Frequency (MHz)	Linux Idle	Linux Max	Windows Idle	Windows Max
600	14.31	16.54	11.24	14.85
800	15.69	20.98		
1000	15.88	22.71		
1200	16.47	25.71		
1300	16.9	27.45	12.84	25.53

Table 5. Calculated values for αC

αC	Linux	Windows
Lowest Idle	9.601 x 10-10	8.18 x 10-10
Highest Idle	1.321 x 10-9	
Lowest Busy	5.578 x 10-9	5.46 x 10-9
Highest Busy	5.052 x 10-9	

obtained are summarized in Figure 1. The backlight current showed no variation with changes in external factors such as the color being displayed on the LCD and the color bit-depth.

The CPU power measurement depended on the formula for power in terms of frequency and voltage, $P_{CPU} = \alpha C V^2 f$, where C is dependent on the capacitance of the chip and α the level of activity. Given two power measurements for P_{CPU}, we can calculate:

$$P_{CPU1} - P_{CPU2} = \alpha C V_1^2 f_1 - \alpha C V_2^2 f_2$$
$$= \alpha C (V_1^2 f_1 - V_2^2 f_2)$$
$$\alpha C = \frac{P_{CPU1} - P_{CPU2}}{V_1^2 f_1 - V_2^2 f_2}$$

Since $P_{CPU} = P_{system} - P_{other}$, holding P_{other} constant implies $\Delta P_{CPU} = \Delta P_{system}$. P_{other} cannot be held exactly constant, but it can be held approximately constant by either leaving the system idle or running a program that stresses only the CPU, without touching main memory or I/O. For this measurement we created a synthetic benchmark that computes several functions and stores the results in an array. The total memory usage of this program is 470KB as measured using top, so it exercises the 1MB cache in the Pentium M, but does not access main memory after an initial phase.

The system power was measured under Linux and Windows, both with the CPU idle and running the cpu-maximizer, at a variety of different CPU frequencies. The results are shown table 4. From these numbers, the value of αC for an idle CPU can be calculated using the formula; the results are shown in table 5. Figure 2 gives the CPU power consumption obtained by using these values of αC. We compared our results to processor power consumption number from Intel [4].

One limitation of this approach is the assumption that α is constant for different applications, which is not necessarily the case.

The remaining components of the stripped system were the memory and graphics chipset. The graphics card power consumption was also obtained using the subtractive method. There were two main problems that we had to deal with. Firstly, ATI has not published the electrical specifications or the power states of the device. Secondly, there was no means of turning off the device or putting it into different power states. Thus, we assumed a base power of 1.09 W based on the values given in the Intel paper [4]. We obtained the active power use by subtracting the system power for the CPU maximizer from the system power for the 3DMark2001 benchmark. The resulting maximum power consumption was 5.1 W.

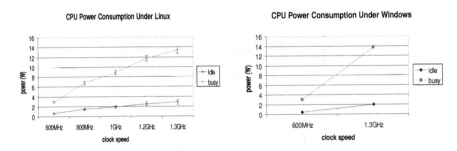

Fig. 2. CPU power consumption under Linux and Windows using calculated αC

The memory power consumption was obtained by the subtractive method in the same manner as the graphics power. We obtained a baseline measure of its power consumption, 0.45 W, from the manufacturer's data sheet [8]. We then obtained system power from a memory test, and subtracted from it the idle system power and the difference in CPU power, yielding 1.42 W. For each of the benchmarks, we computed a memory power figure in between these two by determining the number of memory accesses each benchmark did.

One final power loss we noticed was a baseline 0.65 W loss in the power supply. This was found by measuring the power consumption of the system when nothing is turned on. This figure is just a baseline, and we expect that the true power supply loss under load is substantially higher.

Once the power consumption of each of the components of the stripped system was determined, we proceeded to measure the power consumed by the removable components. Amongst these, we were specifically interested in the power consumption of the wireless LAN card and the optical drive, as the ports on the computer usually consume too small an amount of power to measure accurately with the current probe.

Table 6. Wireless LAN card power consumption

Wireless Card States	Power Consumption
Power Saver (Idle)	0.14 W
Base (Idle)	1.0 W
Transmit	3.12 W total at 4.2 Mb/s
Receive	2.55 W total at 2.9 Mb/s

Table 7. Power consumption of optical drive

Optical drive state	Power (W)
Initial spin up	3.34
Steady spin	2.78
Reading data	5.31

The power consumption of the wireless LAN card was measured using the subtractive method. The wireless card has numerous power states when enabled. To measure these, we measured the system power with the wireless card disabled, enabled but idle, in a power-saving state, while receiving via FTP, and while transmitting via FTP. The power consumption of the system and of the wireless card alone is shown in Table 6. The wireless card power is the difference between the system in the given state minus the system power with the card disabled; for the receive and transmit, we also subtracted the additional power consumed by the CPU and hard drive.

The power consumption of the optical drive was also measured using the subtractive method. First, the system power was measured with the optical drive removed from the system. Then, the system power was measured with the optical drive inserted but idle, while inserting a CD into the optical drive, while the optical drive is spinning steadily, and while the optical drive is reading data. The power consumed by the optical drive in each of these states is the system power in that state minus the system power with no optical drive minus any additional hard drive or CPU usage. Table 7 shows the results.

4   Results

We used several benchmarks described in the experimental setup section to analyze the component-wise breakdown of power consumption. As can be seen in

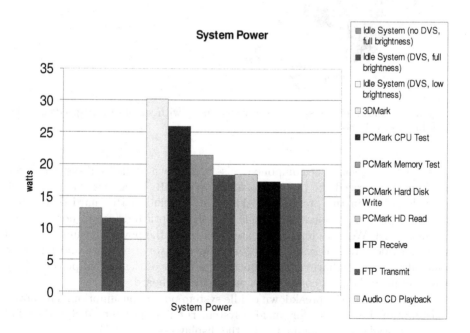

Fig. 3. Power consumption of system under various workloads

Figure 3, the total system power varies by a factor of four depending upon the workload. In the next section we will see the component-wise breakdown of the system wide power, and some of the reasons that contributed to such a large range of total system power consumption.

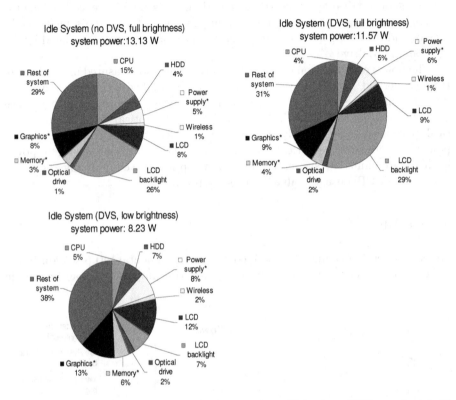

Fig. 4. Breakdown of idle system power conversion with various CPU speeds and display brightness levels

Figures 4 – 8 show the component wise breakdown of the total system power. We have categorized the power consumption into 10 categories, some of which were described in the previous section. These include CPU, hard drive, base power supply loss, wireless card, LCD, backlight, optical drive, memory system, graphics card. We were unable to categorize all of the power consumption, and so we include a rest of system category that consists of memory and interrupt controller hub (Northbridge and Southbridge), rest of the system power supply loss, and miscellaneous components on the motherboard.

Figure 4 shows the breakdown of idle system power consumption. The power consumption breakdown for an idle system running without DVS and at full backlight brightness is dominated by the display system (34%). When the CPU is running at 600 MHz, it uses one-tenth the power used by the display system.

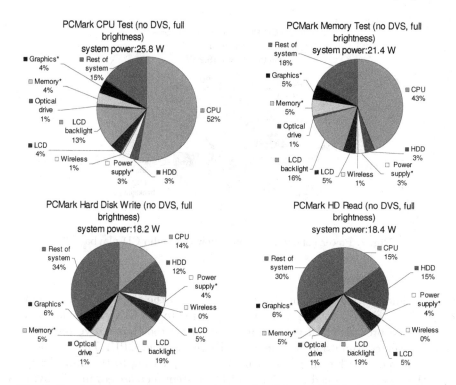

Fig. 5. Breakdown of power consumption under different PCMark2002 benchmarks

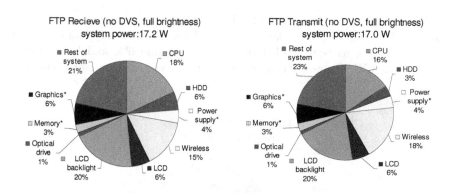

Fig. 6. Power consumption breakdown of FTP

Dimming the backlight does reduce the backlight power, but the display system still accounts for one-fifth of the system power, mainly due to the one-Watt fixed power consumption of the LCD panel.

Next, in Figure 5 we look at the breakdown for PCMark2002. During the course of the CPU tests, the CPU power consumption dominated the total system power consumption. Similarly, in the Memory test, CPU was used a lot for

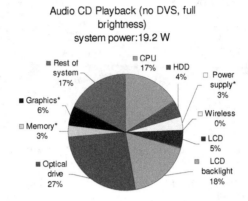

Fig. 7. Power consumption breakdown of audio CD playback

reading, and writing different size blocks. Though the share of memory power consumption went up, it still was small compared to the CPU or Display power. Finally, the breakdown of power consumption for the Hard Drive tests shows equal power consumption by the Hard Drive and the CPU. This is not surprising as most of the time was spent in reading/writing to the disk. Also, we note the power consumed by the rest of the system increases dramatically; this may be because of the increased IDE controller activity that is included in this category.

Figure 6 shows the breakdown for power consumed during a wireless FTP upload and download of the same file. This breakdown is very similar to the Hard Drive tests, as both loaded the CPU by around 6%, and both devices use around 2-3 W of power when active. Transmit was slightly more expensive

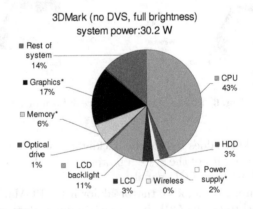

Fig. 8. Power consumption breakdown of 3DMark01

than receive, but comparison is not fair as the transmit speed through TCP was slightly higher than the receive speed.

We can see in Figure 7 that the optical drive uses a lot of power, even more than the processor, during an audio CD playback. This is because the CD was spinning at full speed for the whole duration of the track.

Figure 8 shows the breakdown of power consumption during 3DMark. Not surprisingly, CPU dominates, while graphics comes in a distant second.

5   Discussion

This study provided some useful insights into the power consumption behavior of the individual components, and the manner in which the OS uses these components.

5.1   CPU

This study showed empirical evidence for the fact that DVS saves power, and that these savings are significant enough to contribute to a lower total system power.

A power aware OS can reduce component wise power consumption by exploiting the various low power states supported by a device. In our study we found that the total Idle System power consumption went up when we were running Linux OS compared to the power consumption under Windows OS. This can be attributed to the fact that Linux Kernel we were using did not have support for ACPI, and even after upgrading the kernel we were not able to get ACPI working under Linux. Windows OS on the other hand provides user with a GUI to set the power preferences.

Even though DVS reduces CPU power consumption, CPU still dominates whenever it is extensively used by an application. Using DVS can reduce the CPU power, but lowered frequency and thereby increased execution time may not be acceptable to all users. This implies that there is a lot more room for work in the area of CPU power reduction.

5.2   Hard Drive

One of the interesting observations made during this study was the mysterious Hard Drive accesses made by Windows OS when the drive was in Idle state. These accesses had a frequency of 150 ms and they increased the power consumption of the disk by 0.2 W. Linux OS did not show any such behavior.

5.3   LCD

The relation between LCD power and the color being displayed validates similar findings in adaptive display literature [10]. This also makes a case for light colored screensavers, as the traditional Windows XP screensaver increased the system power consumption by 0.5 W. We also tested the idea of reducing color bit-depth to save power, as proposed in the display system community [10]. We found that for our platform such a scheme did not save any power, either at the LCD panel or the system level. The power reduction that we observed by reducing the backlight brightness level strongly supports automatic brightness reduction as performed by Display Power Saving Technology [11].

5.4   Graphics

From the results, the graphics chipset does not consume a large fraction of the system power except when using 3D acceleration, and even then the graphics chipset only consumes about 17% of the total system power. However, the machine we are measuring contains an older Radeon 7500 chip; newer, faster chips such as the Radeon 9600 may use considerably more power.

Some of the new areas for power savings are certainly at the circuit level, for example the chipset power consumption, and the power supply loss are a major portion of the Idle System power.

5.5   Limitations

The accuracy of our results is limited by several factors. One important limitation is the limits of the power measurement apparatus, particularly the current probe. Agilent's own documentation notes that the probe has a margin of error of +/- 2 mA. Moreover, because the probe works by detecting the magnetic field generated by current flow rather than by directly measuring the current flow, it is susceptible to influence by external magnetic fields. Strong magnetic fields can raise the error margin as high as 20 mA, though the fields in our test environment were fairly small.

Another factor limiting the accuracy of our results was the subtractive method. Because of this method, small errors (¡ 1%) in the system power measurement can translate to very large errors measuring low-power components, as 1% of system power is about 0.15 W. Moreover, because of the inexact methods used to get subtractive measurements of memory, graphics, and the power supply, the measurements in those categories should be regarded as estimates.

A major limitation of this project is that we were unable to get a component-wise breakdown of much of the power consumed by the system. From 14% to 38% of the system power is classified as "rest of platform." Although the information we have discovered is useful, a breakdown of this unclassified power consumption would allow even more insight.

Finally, this project has focused only on the power consumption of the electronic and mechanical parts of the computer, and has completely neglected the idiosyncrasies of the battery. Proper battery management is as crucial to good battery life as reduced power consumption. A future project of this type should also try to assess how much energy the battery wastes under various workloads and drain rates.

6   Related Work

There have been few other studies on breakdown of laptop's power consumption. Intel recently published a paper describing the reduced power consumption of Centrino platform [4][12]. This paper also included a power consumption breakdown, but it does not give much detail on the methodology or the system and the workload that was measured. The breakdown that we obtained for the Idle System with DVS is very similar to the breakdown given in that paper. Another work related to ours is [5], which uses a combination of software profiler and data obtained from industry to come up with a similar breakdown. Although this work is very detailed and informative, it was done a long time ago, and the hardware has since become outdated. Comparing our breakdown numbers to those given in Lorch's paper, we found that the total system power has nearly doubled, but the share of each component has not changed drastically. Our work is different from Lorch's work as we explore the application-dependent nature of a laptop's power consumption whereas his work measures the power consumption under a representative or average set of power states obtained through profiling. There have been several recent studies on power consumption of a hand-held device cignetti [7], but these numbers are not representative of a laptop. Some of the notable differences are that hand-held devices do not have hard drive or graphics card.

7   Conclusions

The purpose of this work was to obtain a component-wise breakdown of the power consumption in a modern laptop. We measured the power usage of the key components in an IBM ThinkPad R40 laptop using an Agilent Oscilloscope and current probes. We obtained the system wide power consumption breakdown for the following components: CPU, optical drive, hard disk, display, graphics card, memory, and wireless card subsystems. Due to the limitations of our measurement equipment, we had to use a combination of direct measurement and subtractive measurement approaches.

We found that total system power consumption varies a lot (8 W - 30 W) depending on the workload. We also found that, although power saving techniques such as DVS can reduce CPU power considerably, the total system power is still dominated by CPU power in the case of CPU intensive benchmarks. The display is the other main source of power consumption in a laptop. We found that reducing the backlight brightness can reduce the system power significantly, more than any other display power saving techniques. The graphics, wireless networking, and disk drives can all consume a substantial amount of power when they are active, but when they are idle, as is the case most of the time, they do not consume too much. Finally, we observed that power consumption under Windows OS differs from power consumption under Linux, probably due to differences in ACPI support.

Last but not the least, we were successful in not destroying the laptop (a fear of all our colleagues).

References

1. Yuan and Nahrstedt. Energy-Efficient Soft Real-Time CPU Scheduling for Mobile Multimedia Systems, SOSP 2003
2. Lebeck, Fan, Zeng, and Ellis. Power Aware Page Allocation, ASPLOS 2000.
3. Weissel, Beutel, and Bellosa. Cooperative I/O - A Novel I/O Semantics for Energy-Aware Applications, OSDI 2002.
4. Chinn, Desai, DiStefano, Ravichandran, and Thakkar. Mobile PC Platforms Enabled with Intel Centrino, Intel Technology Journal, May 2003.
5. Lorch, J. A complete picture of the energy consumption of a portable computer, Masters Thesis, Computer Science, University of California at Berkeley, December 1995.
6. Cignetti, Komarov, and Ellis. Energy Estimation Tools for the Palm, ACM Modeling, Analysis and Simulation of Wireless and Mobile Systems 2000.
7. Flinn, Farkas, and Anderson. Power and Energy Characterization of the Itsy Pocket Computer (Version 1.5), Compaq Western Research Laboratory, Technical Note TN-56, February 2000.
8. Infineon, 256-Mbit DDR SDRAM datasheet, Jan 2003.
9. IBM Mobile Systems, ThinkPad Computer Hardware Maintenance Manual, March 2003
10. Franco Gatti, Andrea Acquaviva, Luca Benini, Bruno Ricco. Low Power Control Techniques For TFT LCD Displays, CASES, 2002.
11. Intel 855 GME chipset mobile computers. http://developer.intel.com/design/chipsets/mobile/855GME.htm
12. Intel Low Power Technologies: Bringing Longer Battery Life and Higher Productivity to Mobile Computing. http://www.intel.com/ebusiness/pdf/prod/related_mobile/wp021601.pdf
13. Intel Pentium M datasheet. http://www.intel.com/design/mobile/datashts/25261203.pdf

Author Index

Lecture Notes in Computer Science

For information about Vols. 1–3712

please contact your bookseller or Springer

Vol. 3762: R. Meersman, Z. Tari, P. Herrero (Eds.), On the Move to Meaningful Internet Systems 2005: OTM 2005 Workshops. XXXI, 1228 pages. 2005.

Vol. 3761: R. Meersman, Z. Tari (Eds.), On the Move to Meaningful Internet Systems 2005: CoopIS, DOA, and ODBASE, Part II. XXVII, 653 pages. 2005.

Vol. 3760: R. Meersman, Z. Tari (Eds.), On the Move to Meaningful Internet Systems 2005: CoopIS, DOA, and ODBASE, Part I. XXVII, 921 pages. 2005.

Vol. 3759: G. Chen, Y. Pan, M. Guo, J. Lu (Eds.), Parallel and Distributed Processing and Applications - ISPA 2005 Workshops. XIII, 669 pages. 2005.

Vol. 3758: Y. Pan, D.-x. Chen, M. Guo, J. Cao, J.J. Dongarra (Eds.), Parallel and Distributed Processing and Applications. XXIII, 1162 pages. 2005.

Vol. 3757: A. Rangarajan, B. Vemuri, A.L. Yuille (Eds.), Energy Minimization Methods in Computer Vision and Pattern Recognition. XII, 666 pages. 2005.

Vol. 3756: J. Cao, W. Nejdl, M. Xu (Eds.), Advanced Parallel Processing Technologies. XIV, 526 pages. 2005.

Vol. 3754: J. Dalmau Royo, G. Hasegawa (Eds.), Management of Multimedia Networks and Services. XII, 384 pages. 2005.

Vol. 3753: O.F. Olsen, L.M.J. Florack, A. Kuijper (Eds.), Deep Structure, Singularities, and Computer Vision. X, 259 pages. 2005.

Vol. 3752: N. Paragios, O. Faugeras, T. Chan, C. Schnörr (Eds.), Variational, Geometric, and Level Set Methods in Computer Vision. XI, 369 pages. 2005.

Vol. 3751: T. Magedanz, E.R. M. Madeira, P. Dini (Eds.), Operations and Management in IP-Based Networks. X, 213 pages. 2005.

Vol. 3750: J.S. Duncan, G. Gerig (Eds.), Medical Image Computing and Computer-Assisted Intervention – MICCAI 2005, Part II. XL, 1018 pages. 2005.

Vol. 3749: J.S. Duncan, G. Gerig (Eds.), Medical Image Computing and Computer-Assisted Intervention – MICCAI 2005, Part I. XXXIX, 942 pages. 2005.

Vol. 3748: A. Hartman, D. Kreische (Eds.), Model Driven Architecture – Foundations and Applications. IX, 349 pages. 2005.

Vol. 3747: C.A. Maziero, J.G. Silva, A.M.S. Andrade, F.M.d. Assis Silva (Eds.), Dependable Computing. XV, 267 pages. 2005.

Vol. 3746: P. Bozanis, E.N. Houstis (Eds.), Advances in Informatics. XIX, 879 pages. 2005.

Vol. 3745: J.L. Oliveira, V. Maojo, F. Martín-Sánchez, A.S. Pereira (Eds.), Biological and Medical Data Analysis. XII, 422 pages. 2005. (Subseries LNBI).

Vol. 3744: T. Magedanz, A. Karmouch, S. Pierre, I. Venieris (Eds.), Mobility Aware Technologies and Applications. XIV, 418 pages. 2005.

Vol. 3742: J. Akiyama, M. Kano, X. Tan (Eds.), Discrete and Computational Geometry. VIII, 213 pages. 2005.

Vol. 3740: T. Srikanthan, J. Xue, C.-H. Chang (Eds.), Advances in Computer Systems Architecture. XVII, 833 pages. 2005.

Vol. 3739: W. Fan, Z.-h. Wu, J. Yang (Eds.), Advances in Web-Age Information Management. XXIV, 930 pages. 2005.

Vol. 3738: V.R. Syrotiuk, E. Chávez (Eds.), Ad-Hoc, Mobile, and Wireless Networks. XI, 360 pages. 2005.

Vol. 3735: A. Hoffmann, H. Motoda, T. Scheffer (Eds.), Discovery Science. XVI, 400 pages. 2005. (Subseries LNAI).

Vol. 3734: S. Jain, H.U. Simon, E. Tomita (Eds.), Algorithmic Learning Theory. XII, 490 pages. 2005. (Subseries LNAI).

Vol. 3733: P. Yolum, T. Güngör, F. Gürgen, C. Özturan (Eds.), Computer and Information Sciences - ISCIS 2005. XXI, 973 pages. 2005.

Vol. 3731: F. Wang (Ed.), Formal Techniques for Networked and Distributed Systems - FORTE 2005. XII, 558 pages. 2005.

Vol. 3729: Y. Gil, E. Motta, V. R. Benjamins, M.A. Musen (Eds.), The Semantic Web – ISWC 2005. XXIII, 1073 pages. 2005.

Vol. 3728: V. Paliouras, J. Vounckx, D. Verkest (Eds.), Integrated Circuit and System Design. XV, 753 pages. 2005.

Vol. 3727: M. Barni, J. Herrera Joancomartí, S. Katzenbeisser, F. Pérez-González (Eds.), Information Hiding. XII, 414 pages. 2005.

Vol. 3726: L.T. Yang, O.F. Rana, B. Di Martino, J.J. Dongarra (Eds.), High Performance Computing and Communications. XXVI, 1116 pages. 2005.

Vol. 3725: D. Borrione, W. Paul (Eds.), Correct Hardware Design and Verification Methods. XII, 412 pages. 2005.

Vol. 3724: P. Fraigniaud (Ed.), Distributed Computing. XIV, 520 pages. 2005.

Vol. 3723: W. Zhao, S. Gong, X. Tang (Eds.), Analysis and Modelling of Faces and Gestures. XI, 4234 pages. 2005.

Vol. 3722: D. Van Hung, M. Wirsing (Eds.), Theoretical Aspects of Computing – ICTAC 2005. XIV, 614 pages. 2005.

Vol. 3721: A.M. Jorge, L. Torgo, P.B. Brazdil, R. Camacho, J. Gama (Eds.), Knowledge Discovery in Databases: PKDD 2005. XXIII, 719 pages. 2005. (Subseries LNAI).

Vol. 3720: J. Gama, R. Camacho, P.B. Brazdil, A.M. Jorge, L. Torgo (Eds.), Machine Learning: ECML 2005. XXIII, 769 pages. 2005. (Subseries LNAI).

Vol. 3719: M. Hobbs, A.M. Goscinski, W. Zhou (Eds.), Distributed and Parallel Computing. XI, 448 pages. 2005.

Vol. 3718: V.G. Ganzha, E.W. Mayr, E.V. Vorozhtsov (Eds.), Computer Algebra in Scientific Computing. XII, 502 pages. 2005.

Vol. 3717: B. Gramlich (Ed.), Frontiers of Combining Systems. X, 321 pages. 2005. (Subseries LNAI).

Vol. 3716: L. Delcambre, C. Kop, H.C. Mayr, J. Mylopoulos, Ó. Pastor (Eds.), Conceptual Modeling – ER 2005. XVI, 498 pages. 2005.

Vol. 3715: E. Dawson, S. Vaudenay (Eds.), Progress in Cryptology – Mycrypt 2005. XI, 329 pages. 2005.

Vol. 3714: H. Obbink, K. Pohl (Eds.), Software Product Lines. XIII, 235 pages. 2005.

Vol. 3713: L.C. Briand, C. Williams (Eds.), Model Driven Engineering Languages and Systems. XV, 722 pages. 2005.